THE CALL
TO
CONVERSION

ALSO BY JIM WALLIS

*God's Politics: Why the Right Gets It Wrong
and the Left Doesn't Get It*

Faith Works

Who Speaks for God?

The Soul of Politics

Revive Us Again: A Sojourner's Story

Agenda for Biblical People

EDITED BY JIM WALLIS

Cloud of Witnesses

Crucible of Fire

Waging Peace

*Peacemakers: Christian Voices from the
New Abolitionist Movement*

*The Rise of Christian Conscience: The Emergence of a Dramatic
Renewal Movement in the Church Today*

THE CALL
TO
CONVERSION

*Why Faith Is Always Personal
But Never Private*

JIM WALLIS

HarperSanFrancisco
A Division of HarperCollins*Publishers*

HarperCollins books may be purchased for educational, business, or sales promotional use. For information please write: Special Markets Department, HarperCollins Publishers, 10 East 53rd Street, New York, NY 10022.

HarperCollins Web site: http://www.harpercollins.com
HarperCollins®, ®, and HarperSanFrancisco™ are
trademarks of HarperCollins Publishers.

FIRST EDITION
Designed by Joseph Rutt

Library of Congress Cataloging-in-Publication Data
Wallis, Jim.
The call to conversion : why faith is always personal but never private /
Jim Wallis. — 1st ed.
p. cm.
Includes bibliographical references.
ISBN-10: 0-06-084237-7
ISBN-13: 978-0-06-084237-6
1. Conversion Christianity. 2. Christian life.
3. Christianity—20th century. I. Title.
BV4916.3W35 2005
248.2'4—dc22 2005052593

05 06 07 08 09 RRD(H) 10 9 8 7 6 5 4 3 2

Contents

Introduction

I JUST FINISHED the book tour for *God's Politics: Why the Right Gets It Wrong and the Left Doesn't Get It*. From the first week onward, book signings turned into town meetings, and bookstore events into revivals. In all we toured for twenty-one weeks, traveling to forty-eight cities, doing eighty-five events, speaking face-to-face to almost seventy-five thousand people, and reaching millions more through hundreds of interviews. A book on faith and, even more amazingly, faith and politics, jumped to number two on Amazon its first week out, and then onto the *New York Times* bestseller list for fifteen weeks and remains on the extended bestseller list as I write this, almost twenty-five weeks from its publication.

Right from the start, I realized something important was happening, and that it was about much more than a book. *God's Politics* became the right book at the right time, and simply revealed what was already there across the nation and was just waiting to be expressed. Many people of faith felt their voice was not being heard in the national debate over faith and politics, and now found something to point to. I soon realized the large numbers of people who came out at virtually every stop were not just coming to hear my voice, but also to express their voice. I'm still amazed how much the national conversation about faith and politics has already changed as a result of the *God's Politics* book and book tour.

The basic message, since I first wrote *Call to Conversion*, has not fundamentally changed but the openness to it in the church, political world, media, and culture has changed dramatically. But more than at any time in thirty-five years, the message has broken through. That is due to many factors: the 2004 election, the heightened role of religion and "moral values" in our political discourse, the tiresome public monologue of the Religious Right, the reaction to the Religious Right's hubris and pursuit of power among a large number of other people of faith, and, perhaps most important, the essential moral and spiritual character of many of the most pressing issues our society confronts—the massive nature of global and even domestic poverty, the crisis of the environment, the cost and consequences of war, the selective moralities of both Left and Right in regard to the sanctity of life, and the breakdown of both family and community.

We now face a new moment of opportunity and possibility and must rise to the occasion—both spiritually and politically. It's instructive to ask who is responding and what are they responding to?

Let's start by asking who is responding. A breakdown of the audiences for the town meetings that *God's Politics* generated includes the following.

First, the audiences have included evangelical Christians who don't feel represented by the ideological, shrill, and divisive rhetoric and behavior of the Religious Right. Many progressive, moderate, and even conservative evangelicals are looking for an alternative to the Religious Right and do not believe that "moral values" can be reduced to one or two contentious social issues and are hungry for a wider and deeper vision of Christian ethics.

Second, many Catholics are coming who don't agree with the instruction of some right-wing bishops who tell them they should vote on only one issue—abortion—and thereby ignore all the rest of Catholic social teaching. That more "consistent ethic" or "seamless garment" of human life appeals to many of them.

Third, Christians from mainline Protestant denominations are well represented and express how disrespected and dismissed they have felt in the debates over faith and politics. Their desire for the integration of a personal faith and social gospel is very evident.

Black Christians are eagerly participating in what they hope will be a new conversation that is not as "white" as the old one, in which the media talk of "evangelicals" only in terms of white evangelicals and consistently ignore the black churches. The more holistic gospel of historically black churches in America has a great deal to offer the wider church. Hispanic and Asian Christians are changing the face of many of our churches and making the audiences in the *God's Politics* gatherings even more diverse. A genuine multicultural future is the vision of all these Christians of color in America.

Many rabbis are also coming, along with their congregations, and we have done events in synagogues. Our ideas and the message of the Hebrew prophets seem to appeal to our Jewish brothers and sisters as an alternative to the Religious Right.

We are also meeting young Muslims who are seeking to forge a more open, tolerant, compassionate, democratic, and peaceful Islam. The internal battle between fundamentalism and prophetic faith is now occurring in all three of the great monotheistic religious—Christianity, Judaism, and Islam. We are also encountering Buddhists and other faith traditions.

And at every event, there are many young people who say they are "not religious but spiritual." And they are very grateful to be included in this moral discourse on our public life as are those who identify themselves as "secular," "agnostic," or even "atheist," but want to be part of the moral values discussion.

The issues that were consistently raised on the book tour and that caused the greatest response include the following.

First, global and domestic poverty was the most unifying and galvanizing issue. A real clarity arose around the nature of a "new altar call" for these times: Make Poverty History (which is the compelling new slogan of the British campaign against global poverty). Overcoming poverty is being seen by many, especially young people, as the natural outcome of faith, even as a test of faith in our time.

Second, the issues of war and peace, and conflict and its resolution, were also a central part of the message and the response on the tour. The critical need for a "moral response to terrorism" and an alternative to unilateral and preemptive war, as in Iraq, became a rallying cry. The observation that conflicts around the world are deeply connected to poverty and are not being resolved by war is increasingly apparent to many.

Third, protecting the environment, otherwise known as God's creation, was also a deeply rooted and growing commitment of those who attended the *God's Politics* town meetings. The alarming crisis of the environment is also socially and spiritually connected to the issues of poverty and war.

Fourth, the partisan manipulation of issues surrounding the sanctity of life is of vital concern to many. Neither the political Left or Right is practicing a "consistent ethic" of life in which all the threats to the dignity and sacredness of life are addressed. A consistent life ethic connects issues such as abortion, euthanasia, capital punishment, war, pandemics such as HIV/AIDS and malaria, and, of course, poverty.

Finally, the breakdown of our closest relationships in family and community is of great concern to many. Parenting is becoming a common-ground issue just as poverty is. Across the religious and political spectrum there are deep and legitimate concerns about the health of marriages, the raising of children, the coarsening of cultural values, and the loss of moral standards. The political manipulation of these legitimate concerns—revealed in controversies about gay marriage, for example—is a sign of how important these issues are to most people. The fraying of bonds of community and the rise of selfish and harsh individualism are also thought to be undermining the common good.

Many of these critical questions were addressed in *Call to Conversion*, originally published nearly twenty-five years ago. Because *Call to Conversion* sought to deeply probe the *theological* basis for some of our most crucial social issues, my publishers at Harper San Francisco thought this would be a good time to revise and republish it. *Call to Conversion* was written to be a manifesto of Christian discipleship, and no topic is more important today. Most of the chapters are substantially the same as in the original version, but updated factually and revised to reflect our social and religious context twenty-five years later. However, the core theology of *Call to Conversion* remains the same after all these years. Chapter four, on the issues of war and peace, needed the most substantial revision. The nuclear arms race at the height of the Cold War was the greatest danger then, but is now being replaced by the dangers of terrorism and the prospect of endless war in response to it. As then, the theological basis of Christian peacemaking remains the same. The final chapter, on resurrection hope, seems as urgent and relevant now as ever. The greatest call is still for our conversion. *Call to Conversion* brings us back to the first call of Christ to his disciples and maps the spiritual and social terrain for conversion in these troubled times. I offer it with a prayer and with the confident hope that conversion is still possible.

July 2005
Washington, D.C.

Introduction to
the 1981 Edition

THE TIMES IN which we live cry out for our conversion. The vast majority of the world's people is virtually imprisoned in poverty while an affluent minority is plagued with anxiety. The earth has been ravaged by the growth of our industrial civilization, and the natural resources upon which we depend are running out. All of us, rich and poor alike, are held hostage to a nuclear arsenal capable of snuffing out millions of lives in minutes.

Everywhere we look, the value of human life seems to be steadily diminishing. A spirit of fearful insecurity and mass resignation abounds. Young people say that they don't expect to live out their lives, and couples hesitate to bring children into the world. We have become alienated from the poor, from the earth, from the survival of future generations, and, at root, from God. The cost of our much-touted style of life has been higher than any of us could have realized. Family life, community spirit, and mutual aid have all been replaced by the lifeboat ethic of protecting, defending, and competing for scarcer resources.

Crisis has become a word to describe our whole way of life. The world appears to be falling apart while social commentators argue the probabilities of which might come first, economic disaster or nuclear destruction. Politicians, unable to change the momentum or face the hard choices, are reduced to boosting national morale while defending the status quo. In this perilous situation, appeals for change based on fear seem only to dig us in deeper. Similarly, individual self-improvement, education, and gradual social reform are all old solutions that seem unable to save us now. Repair in the road is no longer helpful if we are headed in the wrong direction.

Rather, we need to be converted. We need to turn around. We need nothing less than a spiritual transformation. Our need for conversion was made dramatically clear to me one Advent night while writing this book. A flu virus sent me to bed early. Unable to do anything else, I decided to sample an evening's fare of television.

In one night I saw advertised an array of gadgets and comforts beyond the wildest dreams of any previous generation. Products and experiences beyond a king's reach in former times are now offered as Christmas gifts among America's affluent. The evening's fare included automatic garage doors, wristwatches with built-in computer-calculators, electronic toys that simulate battles in outer space, coffeemakers with an alarm system that starts your coffee perking before you get up in the morning, cocktail mixes for before dinner, wines to accompany it, and liqueurs to polish it off, designer jeans to make you sexy, perfumes guaranteed to inflame any passion, and a special one-way trip to Florida reduced to just $99 so that "anyone can afford it" (an absurdity to poor people in neighborhoods like mine). In that one Madison Avenue phrase, poverty was abolished—the poor no longer existed!

All these consumer "goods" were far beyond what any of us could ever need, especially in a world where millions of people find themselves locked in daily battle for mere survival. But this is only one side of our luxury economy. The same companies that make things to delight the rich also make things to defend them. Texas Instruments, a producer of children's electronic games, also makes guided-missile systems aimed at other people's children. General Electric ("We Bring Good Things to Life") manufactures not only those early morning coffeemakers but also the Mark 12-A missile, a first-strike nuclear weapon.

The logic is clear: Our affluence must be protected if we are to control the lion's share of the world's resources and leave a billion people hungry. We cannot create an economy based on overconsumption without creating the weapons necessary to keep the poor masses at bay. Each year, as our affluence reaches new heights, our military arsenal keeps pace. Our military technology rules our political life. Behind the cheery TV commercials lies a quiet, deadly reality.

What was happening through the television that night I watched was spiritual formation. Far more effective than crude totalitarianism, this continual electronic suasion is forming the values, the mind, and the spirit of each of us in our all-consuming society. Such spiritual formation whets our appetite for more while closing our eyes and hardening our hearts to the worldwide consequences of our materialistic way of life. In fact, most advertising appeals directly to one or more of the seven deadly sins: pride, lust, envy, anger, covetousness, gluttony, and sloth. And are the television programs sandwiched around these ads much different? Not really. Even the television preachers tell us that prosperity is evidence of God's blessing; by implication, poverty is a sign of God's disfavor.

How we need to be converted, to be turned around! Only a fundamental transformation in the ways we now think and live can provide any basis for hope. Our energies can no longer be directed to fighting for the lesser evils in our public life. Rather, we must devote ourselves to altering the spirit of the age and the framework of our society. Changes in our basic values and assumptions, both personal and corporate, are urgently needed. But how do we break out of the attitudes and habits that control us? How do we chart a new course?

We must recognize that broken fellowship is the root cause of our personal and political alienation. Our sense of covenant with our neighbor, with the earth, and with God has been lost. We are tragically nearsighted if we fail to see the spiritual connections between our abandonment of the poor, despoiling of the earth, cultural acceptance of racial and sexual exploitation, approval of abortion on demand, and willingness to commit nuclear genocide against enemy populations.

We need a new beginning. In the Bible, new beginnings start with repentance. Repentance is the first step in conversion. Repentance means to face up honestly to the past and to turn from it. Conversion means turning around, going in a different direction, making a fresh start. Old values and habits are left behind as we set our feet on a new path. "For the former things have passed away," says the Lord. "Behold I make all things new" (Rev. 21:4,5).

There are always historical issues that shape the life of the church, and the gospel is most truly at stake in our response to them. When discipleship is made historically specific, it becomes the radical force that turns the world upside down. But first, each generation of believers must decide whether their Christianity will have anything to do with Jesus; if so, then their Christian faith has to do with present historical realities. For Jesus is God breaking into history.

The task of this book is to discover the meaning of conversion in our historical situation. Much of the literature on conversion takes a psychological approach, assessing the relationship between religious experience and the behavioral sciences. Other studies try to be systematic in comprehensively profiling various dimensions of conversion. Our approach here will be to seek the biblical meaning of conversion and to apply it to the particular history we face. The question we seek to answer is: What is the meaning of conversion now?

My own understanding of conversion-in-history has come through years of struggle. Conversion was key in my evangelical upbringing. Like many evangelical children, I was "saved" at a young age. That conversion, however, was rarely related to any concrete historical realities. It remained private and abstract, focused, primarily on a few personal habits and practices.

My conversion began to be historically specific later when, as a teenager, I was confronted with the brutal realities of racism in my hometown of Detroit. I now consider that confrontation the beginning of my deeper conversion. I didn't know it then, however, because I had been taught to expect conversion in well-lit churches, not in dark ghetto streets. The more I learned from my black friends and co-workers in inner-city Detroit, the more I felt disillusioned, hurt, angry, betrayed, and rejected by the church. I became separated from the church and, in the process, I felt I had lost my faith. I found my home instead in the civil rights and antiwar movements of the late sixties.

The church had convinced me that religion was unrelated to present historical realities. Racism was not a religious issue of broken fellowship; it was merely a social issue that had no place in the church. The church would say the same thing about the war in Vietnam—that it was a political issue and not a spiritual question. In retrospect, I see that I was being confronted with racism and war as crucial steps in deepening and concretizing my conversion. I was pushed to make a choice for faith in history, to overcome the separation of faith from the real world.

I was too young to understand how the church I knew was a victim of a terrible split, a tearing of faith into pieces. I couldn't see how the Christians were trapped in fear, in false security, in spiritual paralysis. All I saw—and all the world often sees—is that they didn't care.

After the nationwide student strike of 1970, it became clear to many of us that the end of the student antiwar movement was near, partly because of external pressures but largely as a result of internal collapse. From being able to put ten thousand people in the streets, we dwindled to a small circle of the faithful few. It was a hard time, a time of sorting out, of choosing direction, of deciding how we were going to live.

Some spoke of revolution, but it was short-lived. Most slowly made their way back into the mainstream; the rewards and punishments of the system were simply too strong.

The search for deeper roots led me back to the New Testament. Throughout my student years I could never quite shed myself of Jesus. Now I encountered him as never before. I read the Sermon on the Mount and realized that Jesus was talking about a whole new order. He called me back, or maybe he called me for the first time. I saw in Matthew 25 a king who could be found in the poor and dispossessed, and I knew I wanted to follow him. The old split between personal faith and social action found resolution as I learned more fully what it means to belong to Christ. What it means to belong to Christ today—that became my primary concern.

I had come full circle. Conversion had been the big issue in my evangelical past. I had left the church in order to become involved in what I saw as the real issues of my time. My deeper conversion came in the midst of those questions. I had returned to my roots but was applying my tradition in a new way to new realities.

Since that time I have steadily become more convinced that understanding conversion is really the central issue for today's churches. Conversion understood apart from or outside history must be reappropriated and understood in direct relationship to that history.

The interpretation of the Bible and the discernment of the times offered in this book rises out of the community in which I live—a community reading the Bible in a particular historical situation. In other words, the theology of this book grows out of a context, as does all theology. My context is a Christian community of about fifty people, single, married, and children, who seek a life together. This means sharing with one another all that we have—our resources and gifts, work and play, callings and struggles, joys and pains, hopes and fears. We are a group of very human and ordinary people who have been called to an extraordinary gospel.

My context is also an inner-city neighborhood in Washington, D.C., where a third of the people are out of work, where the poor are losing even their substandard housing, where kids graduate from high school unable to read, where girls bear children when they themselves are still children. It is an environment filled with alcohol, drugs, and violence born of the pervasive belief, especially among the young, that they have no future in this society. Here even little children must become tough to survive.

Finally, my context is the capital city of the richest nation on earth, a city where money and power matter the most and where the poor don't matter at all. Every day, in quiet, carpeted offices, well-dressed officials and bureaucrats with clean consciences make the many necessary plans for nuclear war.

I am a pastor, an editor, a teacher, and an evangelist at heart. I am not a scholar. Therefore this book is more a pastoral letter or a tract for the times than a scholarly work. On matters of faith, it is not systematic theology. On the church, it is not systematic ecclesiology.

This book is addressed primarily to Christians in the United States and in other rich countries. It is intended for clergy and laity, for those involved in evangelical revival and charismatic renewal, for members of mainline churches and Catholic parishes, for political activists looking for faith and vision. We need an authentic Christian voice that can speak with clear relevance to the most crucial questions of our times, that will reappropriate our biblical heritage and apply it to our present circumstances. Why? Because I believe that the power of evangelism, of worship, of prayer, and of community is both stronger and more enduring than the spiritual formation pumped into us by the vehicles of mass culture. Despite certain risk, the relationship of personal faith to public life must be demonstrated anew. This book is directed toward a church that still suffers from its splitting of faith away from history. It hopes to speak to all who struggle with the meaning of conversion for us today. The American emphasis on conversion has often been considered a great strength of our churches. In fact, our understanding of conversion is one of our greatest weaknesses.

Neither evangelicals nor liberals have adequately grasped the meaning of conversion for these times. Both movements are floundering without an understanding of discipleship that is historically relevant. It has been said that evangelicals are strong on evangelism and weak on social action, while for liberals the reverse is true. If each group has half a loaf, the assumed solution is merely a matter of putting the two halves together. Then we would have both evangelism and social action—a "both-and" gospel. But such a pasted-together solution compromises the essential unity of the gospel. We need more faith that is footed in the original message of Christ.

Our call is to seek the conversion of the church in the midst of a crumbling empire, an empire to which the church is now closely allied. Our question is the old question of spiritual formation: How is the mind of Christ formed in us and in our history? To answer that question, we will use an incarnational approach, which begins by taking seriously the following questions: Will we follow Jesus? Is the church to be the presence of Jesus in the world? If so, what would Jesus's presence look like now?

We cannot scare people into conversion ("Things are so bad that . . ."), but we can invite people into conversion ("God so loved the world that . . ."). Only in this way can we reach out to people, especially in the churches, who are confused and anxious about the future.

Our only hope is conversion. All the questions we now face point to that. Anything less will not be enough.

CHAPTER I

The Call

The people who sat in darkness have seen a great light,
and for those who sat in the region and shadow of death
light has dawned. From that time Jesus began to preach,
saying, "Repent, for the kingdom of heaven is at hand."

Matthew 4:16–17

JUST AS LIGHT breaks into the darkness, the kingdom of
God has arrived. That is how the prophet Isaiah, quoted
here by Matthew, said it would be. The times into which
Jesus came were dark indeed. Political domination at the
hands of Rome, economic oppression by the rich, and human
sinfulness on every side—these were the experiences of the
common people. But where there was no light, God's new
order would shine for all to see in the person of Jesus Christ.
No wonder the word *gospel* means "good news"! The people
had been waiting a long time.

Jesus inaugurated a new age, heralded a new order, and
called the people to conversion. "Repent!" he said. Why?
Because the new order of the kingdom is breaking in upon
you and, if you want to be a part of it, you will need to un-
dergo a fundamental transformation. Jesus makes the need for
conversion clear from the beginning. God's new order is so
radically different from everything we are accustomed to that
we must be spiritually remade before we are ready and
equipped to participate in it. In his Gospel, John would
later refer to the change as a "new birth." No aspect of

human existence is safe from this sweeping change—neither the personal, nor the spiritual, social, economic, and political. The kingdom of God has come to change the world and us with it. Our choice is simply whether or not we will offer our allegiance to the kingdom.

> As he walked by the Sea of Galilee, he saw two brothers, Simon who is called Peter and Andrew his brother, casting a net into the sea; for they were fishermen. And he said to them, "Follow me, and I will make you fishers of men." Immediately they left their nets and followed him. And going on from there he saw two other brothers, James the son of Zebedee and John his brother, in the boat with Zebedee their father, mending their nets, and he called them. Immediately they left the boat and their father, and followed him. (Matt. 4:18–22)

Jesus called people to follow him. The first disciples took him quite literally. They were young Jewish men with established occupations and family responsibilities who nevertheless left everything to follow him. Jesus called them to himself, and he called them to a mission. "Follow me, and I will make you fishers of men." Their calling was not just for their own sake. From the outset, Jesus's disciples were—and are—called for a purpose.

To leave their nets was no light choice for these Galilean fishermen. Their fishing nets were their means of livelihood and the symbol of their identity. Now Peter and the others were leaving not only their most valued possessions; they were leaving their former way of life. That is what it meant to follow Jesus. Old ties were broken, former things left behind. Peter said, "Lo, we have left everything and followed you" (Matt. 19:27).

Four simple fishermen heard the call of Jesus. They were the first to obey and follow. They would not be the last. Others too would forsake all previous commitments to join Jesus's band. They would become his disciples and share his life. From then on they were bound to Jesus and to his kingdom; nothing would ever be the same for them again. They had made a clear choice with very real consequences. Jesus told potential converts to count the cost:

> As they were going along the road, a man said to him, "I will follow you wherever you go." And Jesus said to him, "Foxes have holes, and birds of the air have nests; but the Son of man has nowhere to lay his head." To another he said, "Follow me." But he said, "Lord, let me first go and bury my father." But he said to him, "Leave the dead to bury their own dead; but as for you, go and proclaim the kingdom of God." Another said, "I will follow you, Lord; but first let me say farewell to those at my home." Jesus said to him, "No one who puts his hand to the plow and looks back is fit for the kingdom of God." (Luke 9:57–62)

In the Bible, conversion means "turning." "To convert" in the King James Bible is translated "to turn" in the Revised Standard Version.

The Hebrew word for conversion (*shub*) means "to turn, return, bring back, restore." It occurs more than one thousand times and always involves turning from evil and to the Lord.[1] The prophets continually called Israel to turn from its sins and worship of idols and return to Yahweh, the true and living God. This call to conversion was both individual and corporate in the Old Testament. These people of God were much like us, always falling away from their Lord and getting themselves into trouble. Conversion meant to come back, to come home again, to wander no longer in sin, blindness, and idolatry. To convert meant to be again who you really were and to remember to whom you really belonged.

The Greek words for conversion (*metanoein* and *epistrephein*) mean "to turn around."[2] Turning around involves stopping and proceeding in a new direction. The New Testament stresses the necessity of a radical turnabout and invites us to pursue an entirely different course of life. Thus, fundamental change of direction is central to the meaning of the words. The assumption—from the preaching of John the Baptist through Jesus to the first apostles—is that we are on the wrong path, moving away from God. The Bible refers to our self-determined course as walking in sin, darkness, blindness, dullness, sleep, and hardness of heart. To convert is to make an about-face and take a new path.[3]

Correct intellectual belief was a major concern of the Greeks. The early Christians, in contrast, were more concerned with transformation. The first evangelists did not simply ask people what they believed about Jesus; they called upon their listeners to forsake all and to follow him. To embrace his kingdom meant a radical change not only in outlook but in posture, not only in mind but in heart, not only in worldview but in behavior, not only in thoughts but in actions. Conversion for them was more than a changed intellectual position. It was a whole new beginning.

Thus conversion is far more than an emotional release and much more than an intellectual adherence to correct doctrine. It is a basic change in life direction. If the key to conversion in the biblical stories is a turning from and a turning to, it is always appropriate to ask what is being turned from and what is being turned to in the account of any conversion.

Conversion begins with repentance, the Greek word for which is *metanoia*. Our word *repentance* conjures up feelings of being sorry or guilty for something. The biblical meaning is far deeper and richer. In the New Testament usage, repentance is the essential first step to conversion. In the larger rhythm of turning from and turning to, repentance is the turning away from. Repentance turns us from sin, selfishness, darkness, idols, habits, bondages, and demons, both private and public.[4] We turn from all that binds and oppresses us and others, from all the violence and evil in which we are so complicit, from all the false worship that has controlled and corrupted us. Ultimately, repentance is turning from the powers of death. These ominous forces no longer hold us in their grip; they no longer have the last word.

Having begun with repentance, conversion proceeds to faith. The call to repentance is the invitation to freedom and the preparation for faith. Just as John the Baptist prepared the way of Jesus, so repentance makes us ready for faith in Christ. As repentance is the turning from, faith is the turning to. Repentance is seeing our sin and turning from it; faith is seeing Jesus and turning toward him. Together, repentance and faith form the two movements of conversion.[5]

Faith is turning to belief, hope, and trust. As repentance dealt with our past, faith opens up our future. Faith opens us to the future by restoring our sight, softening our hearts, bringing light into our darkness. We are converted to compassion, justice, and peace as we take our stand as citizens of Christ's new order. We see, hear, and feel now as never before. We enter the process of being made sensitive to the values of the new age, the kingdom of God. The victory of Jesus

Christ over the powers of death has now been appropriated to our own lives; we are enabled to live free of their bondage. Christ has vanquished the powers that once held us captive and fearful; we now stand in the radical freedom he bought for us with his own blood. "So if the Son makes you free, you will be free indeed" (John 8:36). Our freedom, like Jesus's, will now become a threat to the existing order of things. It is no mere coincidence that immediately after Jesus says, "You will be free indeed," he says, "Yet you seek to kill me" (John 8:37).

CONVERSION IN THE BIBLE is always firmly grounded in history; it is always addressed to the actual situation in which people find themselves. In other words, biblical conversion is historically specific. People are never called to conversion in a historical vacuum. They turn to God in the midst of concrete historical events, dilemmas, and choices. That turning is always deeply personal, but it is never private. It is never an abstract or theoretical concern; conversion is always a practical issue. Any idea of conversion that is removed from the social and political realities of the day is simply not biblical.

In the biblical narratives, the "from" and "to" of conversion are usually quite clear. Conversion is from sin to salvation, from idols to God, from slavery to freedom, from injustice to justice, from guilt to forgiveness, from lies to truth, from darkness to light, from self to others, from death to life, and much more.[6] Conversion always means to turn to God. But what it means to turn to God is both universal and particular to each historical situation. We are called to respond to God always in the particulars of our own personal, social, and political circumstances. But conversion is also universal: it entails a reversal of the historical givens *whatever they may be* at any place and time—first-century Palestine, sixteenth-century Europe, or the United States in the twenty-first century. As such, conversion will be a scandal to accepted wisdoms, status quos, and oppressive arrangements. Looking back at biblical and saintly conversions, they can appear romantic. But in the present, conversion is more than a promise of all that might be; it is also a threat to all that is. To the guardians of the social order, genuine biblical conversion will seem dangerous.

In both the Old and New Testaments conversion involved a "change of lords."[7] Conversion from idolatry is a constant biblical theme: False gods enter the household of faith; alien deities command an allegiance that rightly belongs to God alone. The people, then as now, resisted the naming of their idols and stubbornly clung to them. What were the idols that lured the people of God? Which were the false gods that demanded service and fidelity? Our contemporary idols are not so different from those of biblical times: wealth, power, pride of self, pride of nation, sex, race, military might, etc. Conversion meant a turning away from the reigning idolatries and turning back to the true worship of the living God.

There are no neutral zones or areas of life left untouched by biblical conversion. It is never solely confined to the inner self, religious consciousness, personal morality, intellectual belief, or political opinion. Conversion in Scripture was not a self-improvement course or a set of guidelines to help people progress down the same road they were already traveling. Conversion was not just added to the life they were already living. The whole of life underwent conversion in the biblical accounts. There were no exceptions, limitations, or restrictions.

If we believe the Bible, every part of our lives belongs to the God who created us and intends to redeem us. No part of us stands apart from God's boundless love; no aspect of our lives remains untouched by the conversion that is God's call and God's gift to us. Biblically, conversion means to surrender ourselves to God in every sphere of human existence: the personal and social, the spiritual and economic, the psychological and political.

Conversion is our fundamental decision in regard to God. It marks nothing less than the ending of the old and the emergence of the new. "When anyone is united to Christ, there is a new world; the old has gone, and a new order has already begun" (2 Cor. 5:17, New English Bible). Heart, mind, and soul, being, thinking, and doing—all are remade in the grace of God's redeeming love. This decision to allow ourselves to be remade, this conversion, is neither a static nor a once-and-finished event. It is both a moment and a process of transformation that deepens and extends through the whole of our lives. Many think conversion is only for nonbelievers, but the Bible sees conversion as also necessary for the erring believer, the lukewarm community of faith, the people of God who have fallen into disobedience and idolatry.[8]

The people of God are those who have been converted to God and to God's purposes in history. They define their lives by their relationship to the Lord. No longer are their lives organized around their own needs or the dictates of the ruling powers. They belong to the Lord and serve God alone. They have identified themselves with the kingdom of God in the world, and the measure of their existence is in doing God's will. Transformed by God's love, the converted experience a change in all their relationships: to God, to their neighbor, to the world, to their possessions, to the poor and dispossessed, to the violence around them, to the idols of their culture, to the false gods of the state, to their friends, and to their enemies. The early church was known for these things. In other words, the early Christians were known for the things their conversion wrought. Their conversion happened in history; and, in history, the fruits of their conversion were made evident.

Biblical conversion is never an ahistorical, metaphysical transaction affecting only God and the particular sinner involved. Conversion happens in individuals in history; it affects history and is affected by history. The biblical accounts of conversion demonstrate that conversion occurs *within* history; it is not something that occurs in a private realm apart from the world and is then *applied* to history.[9]

The goal of biblical conversion is not to save souls apart from history but to bring the kingdom of God into the world with explosive force; it begins with individuals but is for the sake of the world. The more strongly present that goal is, the more genuinely biblical a conversion is. Churches today are tragically split between those who stress conversion but have forgotten its goal, and those who emphasize Christian social action but have forgotten the necessity for conversion. Today's converts need their eyes opened to history as much as today's activists need their spirit opened to conversion. But first, both need to recover the original meaning of conversion to Jesus Christ and to his kingdom. Only then can our painful division be healed and the integrity of the church's proclamation be restored. Only then can we be enabled to move beyond the impasse that has crippled and impoverished the churches for so long.

Conversion in the New Testament can be understood only from the perspective of the kingdom of God. The salvation of individuals and the fulfillment of the kingdom are intimately connected and are linked in the preaching of Jesus and the apostles. The powerful and compelling call to conversion in the Gospels arose directly out of the fact of an inbreaking new order. To be converted to Christ meant to give one's allegiance to the kingdom, to enter into God's purposes for the world expressed in the language of the kingdom. The disciples couldn't have given themselves to Jesus and then ignored the meaning of his kingdom for their lives and the world. Their conversion, like ours, can be understood only from the vantage point of the new age inaugurated in Jesus Christ. They joined him, followed him, transferred their allegiance to him, and, in so doing, became people of the new order. His gospel was the good news of the kingdom of God. There is no other gospel in the New Testament. The arrival of Jesus was the arrival of the kingdom.

Our conversion, then, cannot be an end in itself; it is the first step of entry into the kingdom. Conversion marks the birth of the movement out of a merely private existence into a public consciousness. Conversion is the beginning of active solidarity with the purposes of the kingdom of God in the world. No longer preoccupied with our private lives, we are engaged in a vocation for the world. Our prayer becomes, "Thy kingdom come, thy will be done, on earth as it is in heaven." If we restrict our salvation to only inner concerns, we have yet to enter into its fullness. Turning from ourselves to Jesus identifies us with him in the world. Conversion, then, is to public responsibility— but public responsibility as defined by the kingdom, not by the state. Our own salvation, which began with a personal decision about Jesus Christ, becomes intimately linked with the fulfillment of the kingdom of God. The connection between conversion and the kingdom cannot be emphasized enough.[10]

But what if our particular conversion is misshaped by an inadequate preaching of the gospel or the church's lack of faith? What if our particular conversion event knows little of the intimate connection between conversion and the kingdom? We must submit our conversion to the standard of Scripture. Having stressed the importance of the kingdom, we will now turn to Jesus's own description of it.

The preaching of John the Baptist set the stage for Jesus. His radical call to repentance was clearly in the prophetic tradition, which always called the people of God to return. Jesus's preaching followed directly upon John's call to the people to repent and believe the good news of the kingdom, which he came to announce. Both called for the fundamental turning that is always the substance of conversion. After the announcement of the kingdom, Jesus called his disciples and quickly moved on to set forth the meaning of his kingdom. His description of the new order is found in the Sermon on the Mount (Matt. 5–7; Luke 6). The Sermon explains what the kingdom is all about. Its message is clear and compelling. The Sermon is a practical vision of how to live in the new order and of what it will mean to follow Jesus. It is not a new law, but it is a vivid description of the kind of behavior involved in accepting the good news of the kingdom.

In fact, the Sermon on the Mount is the declaration of the kingdom of God, the charter of the new order. It describes the character, priorities, values, and norms of the new age Jesus came to inaugurate. The early church took it to be a basic teaching on the meaning of the kingdom; the Sermon was used to instruct new converts in the faith.

Examining the content of the Sermon, we quickly realize that this new order is not as theoretical and abstract as we might have hoped. It has to do with very concrete things. Jesus speaks to the basic stuff of human existence. He concerns himself with money, possessions, power, violence, anxiety, sexuality, faith and the law, security, true and false religion, the way we treat our neighbor, and the way we treat our enemies. At stake are not just religious issues. These are the basic questions that every man and woman must come to terms with and make choices about. The way we respond to these issues will determine our allegiance to the kingdom of God.

Yet, the sermon begins not with a list of obligations but with a series of blessings. The beatitudes, as they are called, reveal the heart of Jesus and the core values of the kingdom. Jesus's blessings are for the poor, both in spirit and in substance. They know their need of God. He promises comfort to those who have learned how to weep for the world. Those with a meek and gentle spirit will have the earth for their possession. He blesses both those who are hungry for justice and those who show mercy; they will receive satisfaction and obtain mercy. It is the pure in heart, he says, who will see God. Jesus blesses the peacemakers and says they will be called God's own children. Finally, he blesses all those who suffer unjustly for the cause of right; the kingdom of heaven will be theirs. This is the personality of the kingdom. It is straightforward. It is both gentle and strong.

Jesus goes on to counsel his disciples to live simply and without hypocrisy. He tells them to trust God for their care and security rather than relying on the accumulation of possessions. In Luke's account of the beatitudes, Jesus pronounces a series of "woes" upon the rich and warns of the judgment that awaits them. He tells his followers to turn the other cheek when attacked and to go the extra mile when prevailed upon. Jesus instructs his disciples to love their enemies and not to return evil for evil. If they live this way, they will be like salt and light to the world. If they seek the kingdom of

God first and value it above all else, everything they ever need will be theirs as well.

Jesus concludes by saying that his disciples, like good and bad trees, will be known by the fruit they bear. Not everyone who calls him Lord will enter the kingdom, but only those who obey his words. They will be like the wise person who builds a house solidly, on a good foundation; when the rains come, that person will be prepared. Those who don't listen to Jesus's words will be washed away because their houses are built on sand.

Blessing and cursing in the Bible are matters of life and death. Blessing is life and the power of God poured into our lives. Cursing is, inevitably, to die. Clear in the sermon is the fact that the specific things that Jesus blesses are the very things we most try to avoid. On the other hand, the things that are so opposite to the description of the kingdom are the things we seek most eagerly. We can only conclude that the values of the kingdom of God are utterly incompatible with our own values and the way of the world. Our culture rejects those who live in the way Jesus calls blessed. Only those who are willing to be despised by the world are ready to enter the kingdom of God. The sermon reveals that God's will for us is completely different from our own inclinations and social training.

The kingdom indeed represents a radical reversal for us. Aggrandizement, ambition, and aggression are normal to us and to our society. Money is the measure of respect, and power is the way to success. Competition is the character of most of our relationships, and violence is regularly sanctioned by our culture as the final means to solve our deepest conflicts. The scriptural advice "Be anxious for nothing" challenges the heart of our narcissistic culture, which, in fact, is anxious over everything. To put it mildly, the Sermon on the Mount offers a way of life contrary to what we are accustomed. It overturns our assumptions of what is normal, reasonable, and responsible. To put it more bluntly, the Sermon stands our values on their heads.

Not everyone responded to this upside-down value system the way the fishermen did. The chief priests and scribes were critical and unbelieving from the beginning. These leaders of society, the holders of wealth and power, plotted against Jesus, mocked him, and sought to destroy him. They wielded religious and political authority. Jesus showed no respect or deference toward any of them. Some of his harshest words were reserved for them. He called them "hypocrites" and "vipers"; he referred to his political ruler as a "fox." Jesus's teaching and behavior created conflict with the ruling authorities wherever he went. The kingdom he proclaimed undermined their whole system. His confrontation with the religious and economic powers in the temple was the incident that led to his crucifixion.[11]

To receive his kingdom, Jesus said we had to become as open as children (Mark 10:13–15). Wealth would be a great obstacle (Mark 10:21–25). Pride, self-satisfaction, and complacency would be enemies of his kingdom. Jesus said he came not to save those who already considered themselves righteous, but to call sinners to repentance (Mark 2:17). Humility would be necessary for conversion (Luke 18:10–14).

The Gospels and the book of Acts record examples of many conversions. The theme is constant. The good news of salvation created a changed heart and life in those who heard and received. Whether in Paul's language of being justified by faith instead of works, or in John's picture of passing from darkness to light, the movement is one from death to life.

• • •

THE EARLY CHRISTIANS were referred to as the people of "the Way."[12] There is a lot in a name.

First, it is highly significant that they were called the people of *the Way*. Christians at the beginning were associated with a particular pattern of life. Their faith produced a discernible lifestyle, a way of life, a process of growth visible to all. This different style of living and relating both grew out of their faith and gave testimony to that faith. To all who saw, Christian belief became identified with a certain kind of behavior. Unlike our modern experience, there was an unmistakable Christian lifestyle recognized by believers and nonbelievers alike. That style of life followed the main lines of Jesus's Sermon on the Mount and his other teaching. To believe meant to follow Jesus. There was little doubt in anyone's mind: Christian discipleship revolved around the hub of the kingdom. The faith of these first Christians had clear social results. They became well known as a caring, sharing, and open community that was especially sensitive to the poor and the outcast. Their love for God, for one another, and for the oppressed was central to their reputation. Their refusal to kill, to recognize racial distinctions, or to bow down before the imperial deities was a matter of public knowledge.

Aristides described the Christians to the Roman emperor Hadrian in this way:

> They love one another. They never fail to help widows; they save orphans from those who would hurt them. If they have something they give freely to the man who has nothing; if they see a stranger, they take him home, and are happy, as though he were a real brother. They

> don't consider themselves brothers in the usual sense,
> but brothers instead through the Spirit, in God.[13]

The early Christians were known for the way they lived, not only for what they believed. For them, the two were completely intertwined.[14] The earliest title given to them reflected the importance of their kingdom lifestyle. They were not called the people of "the experience" or the people of "right doctrine" or even the people of "the church." Rather, they were the people of "the Way."

Second, it is equally significant that the Christians were known as *the people* of the Way.[15] More than just individuals who had been converted, they were now a people, a new community of faith, which had embarked together on a new way of life. The first thing Jesus did after announcing the kingdom was to gather a community. To follow Jesus meant to share Jesus's life and to share it with others. From the beginning, the kingdom would be made manifest through a people who shared a common life. Their visible fellowship would be the sign and the firstfruits of God's new order begun in Jesus Christ. Those who had left everything to follow Jesus were given the gift of community with one another. Henceforth they would belong to Jesus and be inextricably bound together as brothers and sisters in the family of God. The call of Jesus was not only to a new commitment; it was also to a new companionship, a new community established by conversion.

The quality of life shared in the Christian community was a vital part of the evangelistic message of the early church. Christian fellowship became the companion of the Christian gospel; demonstration was vitally linked to proclamation. The oneness of word and deed, dramatically evident in their life together, lent power and force to the witness of the early

Christians. In a classic study of evangelism in the early church, Michael Green concludes: "They made the grace of God credible by a society of love and mutual care which astonished the pagans and was recognized as something entirely new. It lent persuasiveness to their claim that the New Age had dawned in Christ."[16] The word was not only announced but seen in the community of those who were giving it flesh.

The message of the kingdom became more than an idea. A new human society had sprung up, and it looked very much like the new order to which the evangelists pointed. Here love was given daily expression; reconciliation was actually occurring. People were no longer divided into Jew and Gentile, slave and free, male and female. In this community the weak were protected, the stranger welcomed. People were healed, and the poor and dispossessed were cared for and found justice. Everything was shared, joy abounded, and ordinary lives were filled with praise. Something was happening among these Christians that no one could deny. It was very exciting. According to Tertullian, people looked at the early Christians and exclaimed, "See how they love one another!"[17] The fervent character of Christian love not only bound them to one another; it also spilled over the boundaries of their own communities and extended to all in need. The economic sharing practiced by the early Christians, together with their generosity toward the poor, was one of the most evangelistic characteristics of their life. Radical, practical love became the key to their public reputation.

The basic movement of conversion is a change of allegiance to the kingdom of God, the good news that Jesus brings. To convert means to commit our lives unreservedly to Jesus Christ, to join his new order, and to enter into the fellowship of the new community. Our sins are forgiven, we are reconciled to God and to our neighbor, and our destiny becomes inextricably bound to the purposes of Christ in the world.

Evangelism is to this end. The purpose of evangelism is to call for conversion and to call for it in its wholeness. The most controversial question at stake in the world, and even in the church, is whether we will follow Jesus and live under the banner of his kingdom. The evangelist asks that question and aims it right at the heart of each individual and at the heartbeat of our society. Evangelism confronts each person with the decisive choice about Jesus and the kingdom, and it challenges the oppression of the old order with the freeing power of a new one. The gospel of the kingdom sparks a fundamental change in every life and is an intrusion into any social order, be it first-century culture or our twentieth-century world. Evangelism that is faithful to the New Testament will never separate the salvation of the individual from visible witness to God's kingdom on earth. Rather, biblical evangelists will show people how to "cast off the works of darkness" and how to live "as in the day" (Rom. 13:12,13), in the light of the kingdom that is coming and has already begun in Christ Jesus.

In every renewal movement since the time of the early church, the true nature of conversion has been freed from the narrow limitations and restrictions imposed by the world, and the wholeness of conversion recovered. The power of evangelism is restored and the gospel again becomes a message that turns things upside down. The task of the evangelist is not to make the gospel easy but to make it clear. Instead of merely passing on knowledge or imparting an experience, evangelism should call for (and expect) a radical change in behavior and lifestyle.

The unequivocal assertion of the evangelist is that we are saved only through Jesus Christ. Evangelism refutes every ideological prescription for the salvation of the world, defying the suggestion that we can, after all, save ourselves.

The recovery of the fullness and centrality of conversion is essential to genuine renewal. The monastic movements of the Middle Ages, the radical reformation of the sixteenth century, and the evangelical revivals in eighteenth-century England and nineteenth-century America were each marked by a primary emphasis on conversion. That emphasis continues today in the revolutionary consciousness of Third World Christians. Gustavo Gutierrez calls conversion "the touchstone of all spirituality."[18]

Our need, in the rich countries of the northern hemisphere, is for a fresh consciousness of conversion. In the midst of social conditions so oppressive to others and to ourselves, we must again turn to Jesus. Then will authentic evangelism flower and genuine revival break forth in this land once more. But first we must examine and honestly face up to the ways our evangelism has been corrupted and our conversion distorted.

CHAPTER 2

The Betrayal

But in your hearts reverence Christ as Lord. Always be prepared to make a defense to any one who calls you to account for the hope that is in you, yet do it with gentleness and reverence.

1 Peter 3:15

I REMEMBER A conference in New York City. The topic was social justice. Assembled for the meeting were theologians, pastors, priests, nuns, and lay church leaders. At one point a Native American stood up, looked out over the mostly white audience, and said, "Regardless of what the New Testament says, most Christians are materialists with no experience of the Spirit. Regardless of what the New Testament says, most Christians are individualists with no real experience of community." He paused for a moment and then continued: "Let's pretend that you were all Christians. If you were Christians, you would no longer accumulate. You would share everything you had. You would actually love one another. And you would treat each other as if you were family." His eyes were piercing as he asked, "Why don't you do that? Why don't you live that way?"

There was more sophisticated theological and political analysis per square foot in that room than most places. Yet no one could give an answer to the man's questions. He had put his finger on the central problem we face in the churches today. Our Scriptures, confessions, and creeds are all very public, out in the open. Anyone can easily learn what it is supposed to mean to be a Christian. Our Bible is open to public examination; so is the church's life. That is our problem. People can read what our Scriptures say, and they can see how Christians live. The gulf between the two has created an enormous credibility gap.

The evangelism of the church has no power when the essence of the gospel is not lived out in the world. Peter, writing to the early Christians, said, "Always be prepared to make a defense to any one who calls you to account for the hope that is in you" (1 Pet. 3:15). Which is to say, always be ready to explain yourself.

When Peter told the early Christians to be prepared to answer for their faith, he was making an assumption that we dare not miss. He assumed that certain questions would be asked of Christians: "Why do you people live the way you do? It's a mystery to us. It's contrary to our whole way of life. So what motivates you?" Peter, realizing that such questions would be asked, wanted Jesus's followers to be prepared for them. He told them to be ready, when the questions were asked, to give reason for the hope that was within them.

The power of today's evangelism today is tested by the question, What do we have to explain to the world about the way that we live? But that question is no longer being asked of Christians. No one is asking why we live the way we do. Why? Because most people already know the answer: Christians often live the way they do for the same reasons that everybody else lives the way they do. The life of North American churches has become utterly predictable on sociological grounds. Factors of race, class, sex, and national identity shape and define the lives of Christians just like everybody else. No one expects anything different of Christians. The predictability of the Christian style of life, or, more to the point, the loss of a distinctively Christian lifestyle, has severely damaged our proclamation of the gospel. We have lost that visible style of life that was evident in the early Christian communities and that gave their evangelism its compelling power and authority.

Evangelism in our day has largely become a packaged production, a mass-marketed experience in which evangelists strain to answer that question that nobody is asking. Modern evangelists must go through endless contortions to convince people that they are missing something that Christians have. Without the visible witness of a distinct style of life, evangelists must become aggressive and gimmicky, their methods reduced to salesmanship and showmanship. Evangelism often becomes a special activity awkwardly conducted in noisy football stadiums or flashy TV studios, instead of being a simple testimony rising out of a community whose life together invites questions from the surrounding society. When the life of the church no longer raises any questions, evangelism degenerates. The phrases "Jesus saves" and "Jesus is Lord" have been so often used in the absence of any visible historical application that most people simply do not know what the words mean anymore. Perhaps never before has Jesus's name been more frequently mentioned and the content of his life and teaching so thoroughly ignored. One wonders what remains of Jesus in American culture except his name. Unless accompanied by a clear demonstration of what it means to follow Jesus now, the evangelistic message falls on deaf ears or, even worse, calls people to a conversion empty of real content.

The betrayal of the biblical call to conversion has occurred across the theological spectrum.

The gospel must be preached in context. We live in one of the most self-centered cultures in history. Our economic system is the social rationalization of personal selfishness. Self-fulfillment and individual advancement have become our chief goals. The leading question of the times is, "How can I be happy and satisfied?"

Not surprisingly, our self-centered culture has produced a self-centered religion. Preoccupation with self dominates the spirit of the age and shapes the character of religion. Modern evangelism has played right along with this central theme. The most common question in evangelism today is, "What can Jesus do for me?" In other words, the question is how Jesus can help us make it in the present order, not how we can respond to the new order. Potential converts are told that Jesus can make them happier, more self-satisfied, better adjusted, and more prosperous. Jesus quickly becomes the supreme product, attractively packaged and aggressively sold to a consuming public. Complete with billboards, buttons, and bumper stickers, modern evangelistic campaigns advertise Jesus in a competitive market. Even better than Coca-Cola, Jesus is "the Real Thing."

The gospel message has been molded to suit an increasingly narcissistic culture. Conversion is proclaimed as the road to self-realization. Whether through evangelical piety or liberal therapy, the role of religion is presented as a way to help us uncover our human potential—our potential for personal, social, and business success, that is. Modern conversion brings Jesus into our lives rather than bringing us into his. We are told Jesus is here to help us to do better that which we are already doing. Jesus doesn't change our lives, he improves them. Conversion is just for ourselves, not for the world. We ask how Jesus can fulfill our lives, not how we might serve his kingdom.

An insidious characteristic of narcissism is that it causes the loss of a sense of the past and of the future.[1] In the extreme, there is no history to draw from and no concern for future generations. There is only me and now. My satisfaction today is the only important thing. So it is with narcissistic conversion. The richness of the history of the people of God is lost, as is the future of the kingdom. The central faith experience becomes focused on how God is meeting our needs here and now. Our prayers are not for peace but for parking spaces. We lose all consciousness of participating in the purposes of God, which stretch us back in time and move us ahead to a definite future. Our solidarity with the historic community of faith is lost, and the only relationship to the future is to keep pursuing the American dream while, secondarily, waiting for Jesus to come again. But when conversion is devoid of past and future, it is also emptied of any gospel meaning in the present.

I once thought that the gulf between what the Scriptures say and how Christians live was simply the result of self-interest and hypocrisy. There are enough examples of both in the churches today to make a strong case for that thought. But I no longer believe that either self-interest or hypocrisy is the root cause of the great contradictions in the church's life. They have more to do with lack of faith. Our communion with God and with one another is so small that we just do not have the strength or the resources to live the way Jesus taught.

The American church's two greatest afflictions are spiritual lukewarmness and political conformity. Both grow from a lack of faith. And each causes and feeds on the other. Our shallow faith easily acquiesces to the system, and our accommodation to the political order creates an empty spirituality. We have forgotten what conversion means. One sign of that forgetfulness is that despite the millions of evangelical converts, the symptoms of a deep spiritual malaise across this land have not been relieved. In fact, those symptoms include the

Christian preoccupations with money, power, and success. At bottom, our conformity to the world about us is due to a lack of faith. That lack is pervasive enough in the United States in the 1980s to be called a crisis of faith.

We must first understand the nature of our predicament if we hope to find our way out of it. A pattern characteristic of the life of God's people is evident in the Scriptures. It can be seen in the histories of both Israel and the church, and perhaps it helps to explain our present situation. That pattern is a cycle of faith and faithlessness marked by three stages. It begins when God's people forget who they are and to whom they belong. Having forgotten, they soon fall into idolatry. Finally the idols are named, and the people are called back to the Lord.

The Bible and the history of the church reveal that our tradition is one of very forgetful people. We easily lose our memory and our identity as God's children. Uncertain of who we are, we become easy prey to forces from the surrounding culture. The power of those outside influences grows stronger than anything happening within the community of faith. In biblical language, we become vulnerable to false gods and fall into idolatry, which is the second stage of the cycle.

With no strength to resist the idols that dominate our culture, God's people fall away. Eventually, they do not even see the need to resist; rather, they find ways to make their religion compatible with the worship of the other gods. The Israelites usually didn't reject the worship of Yahweh altogether; they wanted to worship Yahweh *and* Baal. Like the people around them, they were loyal to many gods.

The same is true today. Our churches do not dispense with the worship of the Lord; they simply include the worship of other gods. We want God's life, but we want the good life too. We seem to believe that we can pay homage to our many cultural idols and still retain our integrity as God's people.

The Israelites were not allowed such behavior. The prophets railed against the worship of many gods; they continually reminded the people of the first commandment: "I am the Lord your God, who brought you out of the land of Egypt, out of the house of bondage. You shall have no other gods before me" (Deut. 5:6–7). The idols that had crept into the household of faith and set themselves up as rivals to the Lord were rebuked. The prophets didn't just call for righteousness and justice in general; they were specific and named names ("It is you, O King"). The idols that had captured the hearts of the people of God were unmasked and their power destroyed, just as Elijah challenged and defeated the prophets of Baal (1 Kings 18:17f.). This marked the third stage of the cycle.

But the prophets did not simply denounce and indict. Their vocation was undertaken out of love for the people and a holy desire to see them restored to the Lord. The prophetic task was twofold: to name the idols and to call the people back to the Lord. In order to free the captives, the captivity had to be named. The prophets pointed the way of return to God by restoring the collective memory of the people.

I believe the American churches are in the midst of that same cycle. We have forgotten who we are as God's people, and we have fallen into the worship of American gods. Now God's word to us is to return. Church historians may someday describe our period as the "American Captivity of the Church." It is no less real than the Babylonian Captivity in the history of Israel. Trapped in our false worship, we no longer experience the freedom that is our birthright in Jesus Christ. We are subject to alien deities whose influence is greater than anything occurring in our local congregations. Our need is for conversion, for a rekindling of the memory of who we are and for a return to our first love.

We have seen that Jesus's first sermon was a simple one: "Repent, for the kingdom of heaven is at hand." That kind of preaching is little evident in the churches today. In the U.S. churches, it is not the kingdom of God that is at hand; it is the American culture that is at hand. It is the social, economic, and military system of the United States, not the kingdom of God as reflected in the Sermon on the Mount, that reigns supreme. This self-evident fact derives from a failure of conversion and has become the principal obstacle to genuine faith in our time. Our conformity to the culture has made the fullness of the teaching of Jesus incomprehensible to many. Our conformity has left our congregations emotionally high but spiritually weak.

A good evangelist will normally devote the first half of a sermon to sin and the second half to salvation. The evangelist makes known the reality of sin and its consequences, heightening the people's consciousness of wrongdoing and deepening their awareness of God's judgment on evil. But then, the preacher calls them to repentance, the promise of forgiveness, and the offer of new life in Christ. It's a familiar and time-honored pattern. Evangelism is to reveal the fact of sin and to show how Jesus is the answer to it.

One can tell a great deal about evangelists by listening to their definitions of sin and salvation. If Jesus is the answer, what are the questions? What are the things the evangelists consider to be sinful? What are the most important issues? And how is Jesus the answer? In their preoccupation with individual salvation, twentieth-century American evangelists very seldom pointed to our national values or institutions as evidence of sin. Sin is located only in the individual heart, not in the economic system. We don't hear much preaching about how the way of Christ would undermine the practice of American racism, capitalism, or militarism. The U.S. evangelists of our era have been remarkably silent about those places where the gospel of Jesus Christ plainly contradicts the cultural consensus. Armed with a largely personal definition of sin, modern evangelists lost the capacity to relate the gospel to the collective evils of our times.

Ironically, the more successful modern evangelism has become, the less able it has been to communicate the relevance of Jesus's life to this society. Mass evangelism, in particular, has succumbed to the temptation to make the gospel palatable. The gospel must be preached to all, but we cannot sacrifice its radical demands for the sake of mass acceptance. Registering "decisions for Christ" is simply not enough. Before we ask people to make a commitment, we must first dare to inform them of what it *really means* to follow Jesus. The cost of discipleship must be made known in our wealthy and powerful nation. To offer forgiveness of sins but leave out the message of the kingdom is to be unfaithful to the gospel. A gospel of easy belief and simple formulas is not the message of the New Testament. Conversion does bring release from anxiety and deliverance from personal sin. But in the New Testament, that is not the whole of it.

Our problems finally are due to the fact that Jesus, obscured in the American culture, has become obscure even in the churches. For all the invoking of his name, Jesus's presence remains hidden. Many Americans, including many Christians, have little concrete understanding of Jesus, especially in the facts of his earthly life. The historical character of Jesus of Nazareth is quite unknown, while a heavenly Jesus is proclaimed as our Savior. This is a shocking reality in a country where Bibles are perennial best-sellers. Americans not only see the Bible everywhere they turn, even in motel rooms; they also probably refer to it as much as any people in history. Yet the meaning of Jesus's birth, life, teaching, death, and resurrection is clouded. When all we know is that he "saved us from our sins," we cannot see the vision of the new kingdom he brought us and paid his life for.

Biblical scholars will seldom deny the radical thrust of the Sermon on the Mount, and they agree that Jesus taught such things. But many theologians find innumerable ways to moderate and relativize the teaching of Jesus and, in some cases, to set it aside altogether. The issue is not generally over what Jesus really said but whether his words should be normative for us. Evangelical Christians have a particular problem here because of their high Christology, their view of Jesus as the supreme revelation of God. The central tenet of evangelical faith is the absolute authority of Jesus Christ. He is both Savior and Lord. Most theological attempts to moderate or circumvent the teaching of Jesus, however, are based on a low Christology. The authority of Jesus is diminished or restricted to only particular areas of life.

Tragically, and not without some painful awkwardness, today's evangelicals are walking a precarious tightrope between these two conflicting views of Jesus. By training, Jesus is Lord; his teaching must carry absolute authority in our lives. Yet by experience, evangelicals have accepted the ethical conclusions of theologies that have a low view of Jesus's authority. This conflict is at the heart of the problem of present-day evangelism.

Throughout too much of evangelical training and experience there is no clear proclamation of the kingdom of God.

That is the single greatest weakness of evangelistic preaching today. By neglecting the kingdom of God in our preaching, we have lost the integrating and central core of the gospel. The disastrous result is "saved" individuals who comfortably fit into the old order while the new order goes unannounced. The social meaning of conversion is lost, and a privatized gospel supports the status quo. This fundamental distortion of the gospel serves well the interests of wealth and power. Listening to many evangelistic preachers today, one might never know that the coming of Jesus was intended to turn the world upside down.

But we *do* know that, or we *should* know it, because we know of the integral connection between conversion and the kingdom. Orlando Costas comments on the ancient connection and the modern separation: "Conversion has a definite 'what for?' Its goal is not to provide a series of 'emotional trips' or the assimilation of a body of doctrines, nor to recruit women or men for the church, but rather to put them at the service of the mission of God's kingdom."[2]

The mission of God's kingdom cannot be served if the proclamation of the kingdom is not heard. As long as modern evangelism insists on reducing the gospel message to one of personal salvation, that proclamation cannot be heard in its fullness and richness. Our privatizing of the gospel has suppressed the kingdom; and our suppressing of the kingdom has privatized the gospel. Without the kingdom, the gospel is stripped of its public meaning.

The Scriptures teach that evil is rooted not only in the human heart but also in the principalities and powers, in the structures of society.[3] According to the Bible, social sin is often accompanied by an inability to recognize the sin, described as blindness. Often, we are involved in destructive social arrangements without being aware of it. We are barely conscious of the harm we inflict on others when it is done through the social institutions to which we belong. Personal sin is more visible to us than sin rooted in the system. Gregory Baum suggests that our infidelity to God in social sin is rooted in false consciousness.[4] Like an illness, it destroys us while we are unable to recognize its features or escape its power. The powerful ideologies set up to justify and defend social systems have a strong grip on our lives. We cling tenaciously to the beliefs and symbols that make our institutions seem right and good, and we easily overlook the sin built into the system, even as it destroys others' lives and eats away at our own humanity. The slave trade, institutional racism, the inequitable division of the world's wealth, the Nazi horror, the oppression of women, or the nuclear arms race—each exemplifies a blindness that inevitably leads to hardness of heart. The prophets punctured such collective myths and delusions. They called the people to see their disobedience to God and the harm they were doing to others through the structures of their corporate life.

Few American myths and delusions are being punctured today; when they are, it isn't done by most evangelical preachers. The destructive behavior of economic and political structures is not generally a subject of evangelistic sermons. A sole emphasis on Jesus as personal Savior can, and has, led to a defense of the status quo. In the name of Jesus, our blindness increases. What a terrible reversal of the original gospel message! The reversal is so complete, the blindness so total, that today wealthy and powerful interests actually use evangelism to focus people's attention on their personal sins and to distract their attention from the reality of exploitation and oppression.

Evangelism must recover the social meaning of sin and salvation. Our preaching has to make us newly aware of our active and complicit involvement in what the Bible describes as "the sin of the world." That same preaching has to create a new awareness of the kingdom of God.

To reveal the collective evil in which we participate is clearly a part of the evangelistic task. To turn from our social sin is part of our conversion. Genuine evangelism will spark repentance not only for our personal histories but also for our collective histories. We repent for both the wayward path of our personal lives and the wrong direction of our corporate life. To convert to Jesus Christ is to rise above both personal ego and cultural blindness.

I remember the story of Zaccheus from Sunday school. I only recall being taught that Zaccheus was too short to see Jesus, so he had to climb a tree. Now, that is not the real point of the story. The significance of the story of Zaccheus is that he was converted to Jesus and immediately made reparations to the poor. He acted to restore justice to those he had wronged in the exercise of his occupation. Jesus had high praises for him. Zaccheus had recognized his social sin, turned from it, and sought to repair the damage he had done. The conversion of Zaccheus is a paradigm for rich Christians in the world today.

Regaining the full personal and social meaning of conversion is essential. The evangelistic task before us is to make Jesus historically visible once again. To do that, we will have to restore the message of the kingdom of God to our evangelistic proclamations. We need the kind of preaching that will make the presence of Jesus known in the midst of a culture that pays him lip service but is hostile to his message. Such preaching would enable the church to clarify its allegiance and its identity. Christians could once again learn how to live a life that is not only different from the world, but different in ways that really matter. We might even transcend the legalistic separations of the past and the cultural conformity of the present. We could change the church's identity from being just a religious version of the established order. Christians could begin to live, work, play, raise their children, build community, and act publicly in a new and different way that testifies to the vitality of the life of Christ among them.

The questions now emerging in our history will test the depth and integrity of our conversion. As always, our response to concrete historical realities will show whether or not we really belong to Christ. In the following two chapters we will explore two central questions, the economic division of the world into rich and poor and the growing threats of violence. On each question we will attempt to look at the realities, to draw from the biblical material on the subject, and to explore what conversion means for us in relationship to the issue. It is my conviction that our response to these two questions will greatly determine if and how biblical conversion can be restored in our times.

A few words of caution might be in order first. The discussion of economics and of violence was written, and is meant to be read, in the context of the chapters that precede and follow it. Conversion means more than just plopping a concern for justice or peace into the church's life and expecting it to take root. It won't. The church's whole way of life must be converted, and the foundations have to change. The specificity of the next two chapters is thoroughly intentional. But the specificity means that we won't be talking about other issues and assumptions in the church's life that also need transformation.

Our response to these critical issues will determine whether the twenty-first-century church, including evangelicals, can transcend its twentieth-century captivity.

CHAPTER 3

The Injustice

But if we have food and clothing, with these we shall be content. But those who desire to be rich fall into temptation, into a snare, into many senseless and hurtful desires that plunge men into ruin and destruction. For the love of money is the root of all evils.

1 Timothy 6:8–10

M Y SISTER BARB, who was part of our community in inner-city Washington, was headed down one of our local side streets with her two little boys on their way to the neighborhood daycare center that we operate. Three-year-old Michael was intently surveying the scene on the block. Finally, he looked up and said, "Mommy, there was a war here, wasn't there?"

Some of our neighborhood still looked that way. It was thirteen years since the burning and rioting that followed the death of Martin Luther King Jr. Since then this area of Washington, D.C., was known as the "14th Street riot corridor." The signs of urban cancer were still evident in run-down houses, abandoned buildings, and vacant lots. Broken glass and garbage covered the neighborhood like a big dirty carpet. Apartments were overcrowded. So were the welfare rolls. Joblessness and crime rates climbed together.

There were the children, many not yet old enough for school, out on the streets, running, fighting, and getting into trouble late into the night. Alcohol and drugs became a way of life here, necessary pacifiers of the rage, bitter frustration, and despair that poverty and powerlessness create. For a long time, everything was inadequate in the neighborhood: education, health care, sanitation, police protection, and housing. The one thing not lacking was the resiliency of the people, their character shaped by the struggle to survive, enduring these subhuman conditions and making out the best they could under the circumstances.

The people in this neighborhood suffered the same fate as the urban poor in so many other vast inner-city territories around the country: a combination of economic exploitation and political abandonment. They waited patiently and impatiently for the promises made them to be kept. They are still waiting. They have been the objects of political abuse and the subjects of political rhetoric, but the conditions of their lives have not improved. In fact, things have gotten steadily worse.

In the past few years, however, something else has begun to happen in our neighborhood of Columbia Heights, something very basic and far-reaching. The people of the neighborhood see it and know what it means for them. And they are deeply afraid. Their new fear is the fear of displacement. All through the nation's capital, the poor face the imminent threat of being driven from their homes with no place to go. Many have already been made homeless; others wait to become part of the future statistics indicating massive dislocation of low-income people from the inner city.

Among the gutted shells, deteriorated apartment buildings, and overcrowded row houses, a new addition to the Columbia Heights landscape is appearing: beautifully renovated, single-family houses for new residents, who are mostly white and largely from the suburbs. The poor are still here, but the middle class has now arrived. Our neighborhood is now the focus of real estate speculation. The media calls this phenomenon the "back-to-the-city movement" or "gentrification." It simply means the movement of higher-income residents back into older city neighborhoods. And it means the displacement of lower-income residents from those neighborhoods. For all parties effecting this urban transition, profitability is the key. When the bottom line is profit-maximization, and when housing becomes just another commodity to be bought and sold in the marketplace, the needs of people are subordinated.

The choice seems to be between continued urban blight that oppresses the poor, or urban development that displaces them. First the poor are ignored and abandoned in the inner cities. Then the poor are ignored and abandoned again as development displaces them. The desirability of living in the city or in the suburbs may change, but nothing changes the way the poor are pushed around. Like our historic treatment of Native Americans and their land, we give the poor and racial minorities what we don't want, then change our minds and take it back when it becomes desirable. Federal budget debates offer further proof, if it were needed, that the poor don't count in this society. One frustrated congressman said, "Hell, anybody can vote for a cut in legal services for the poor. There's no political problem in voting to kick the hell out of poor people."[1]

I remember the riots that erupted in Detroit during the summer of 1967. Working there for the summer, I felt the terror of a city at war, saw the devastation, and listened to the anger and despair of black friends and co-workers. A spirit of rebellion characterized the young blacks in the Detroit ghetto who were my peers. The response of the police was unrestrained brutality that knew no bounds. The impact of that summer on me went very deep and has stayed with me ever since.

Massive outbreaks of urban violence seem to be required for America to rediscover its poor in the nation's inner cities. Politicians, sociologists, and journalists make headlines with social statistics and analysis intended to explain what happened and why. What the experts always find (and what anyone who has lived or worked with poor people already knows) is that the major indices of poverty, unemployment, and crime show that the plight of the urban poor steadily worsens in many urban areas. Recommendations are made and then forgotten—a pattern that has become a way of life in this country. The poor have never been the objects of more exhaustive analysis while being the victims of such callous neglect.

The maximum amount of money ever spent in one year for the entire poverty program of the 1960s equaled the cost of three weeks of the Vietnam War. While it has become politically chic to criticize the failure of government spending to solve social problems, the true blame for the dearth of public resources for the poor must be placed with an endlessly escalating military budget. In the meantime, through tax subsidies and other transfer payments, the use of public money to support major corporations and the affluent has always far exceeded all government assistance to the poor. This was the biggest scandal of "the welfare state"—a system that allocated most of its resources to make war and subsidize the rich, leaving the poor with precious little except the bum rap of being called "welfare cheaters." When the government tightens its belt, it tightens it around the necks of the poor. Services were cut in the ghettoes, and restrictions added that further dehumanized welfare recipients, most of whom would prefer the dignity of jobs.

The existence of the new, black middle class has proven attractive to the media and useful to whites generally. It allows most Americans to feel that racial problems are largely solved; it enables the white society to ignore the majority of the black population still trapped in poverty and discrimination. Even many of the old liberal allies of the civil rights movement have long since moved on to other issues. The impression propagated through television, magazines, and movies that blacks are "making it" serves to increase the anger and self-hate of those who know that they aren't. They are still shut out of the "good life," which is daily flashed before their eyes on the screen of that one modern appliance that most poor families own.

Unfortunately, many blacks who have achieved middle-class status have left the poor black community behind, moving on to better jobs and better neighborhoods. This has created a vacuum of leadership and role models that is often filled by pimps, hustlers, and dope dealers; they become the new symbols of success for the young. An entire class of mostly young and nonwhite urban dwellers has been left behind, stuck in a permanent cycle of poverty and racism. Black and Hispanic youth form the core of this urban underclass, a whole generation that is in great danger of simply being lost to unemployment, violence, and crime. Despite the promise that this society associates with youth, they have no future in this country; they know it, and they feel less and less stake in the society and its established rules.

One black store owner, who had built a good business from nothing, was wiped out in one night of rioting in 1977 in New York City. He was hardly in a mood to justify the looting, but he said, "The people have been looted too. You have to look at the total economic condition of the people. Window shoppers finally got a chance to fulfill their desires and not just live with the bare necessities. Everybody stepped into the television commercials for a few hours and took what they wanted."

Some people charged that speaking of the poverty of the looters justified the looting. They missed the point. The looting *by* the poor simply mirrors the looting *of* the poor that has gone on in this country for a long time. The spirit of pillage in our inner cities is a crude and desperate reflection of an entire economic order whose basic premise is looting—the looting of the poor by the rich, legitimized and rationalized into a global economic system. Both the urban looters and the corporate managers of this society share a common reverence for the supreme value of money and what money can buy. Their method, at root, is even the same—selfish acquisition, using whatever violence may be necessary. To put it more biblically, the worship of mammon holds sway both in the corporate boardrooms and on the streets; the only difference is in the accepted rules of the game and in the degree of success.

The urban underclass has been made economically expendable. Its unemployment and impoverishment is regarded as an acceptable cost of the system. A cartoon in the *Philadelphia Inquirer* put it well. A woman standing in the living room of a highrise apartment is looking out a picture window at the smoldering inner city some distance away. In her hand is a newspaper; on her face, a look of concern. Her husband, reading a magazine and looking in the other direction, says, "The problem of the ghettoes? The ghettoes, my dear, are a solution, not a problem."

Blaming the poor for their poverty is one of our most venerable American traditions, and one of our most cruel. Urban rioting was merely the underside of American consumerism, but America attacked the rioters anyway. While it may be an

American tradition to blame the victim, it clearly is not a biblical one. A more biblical reading of our situation would suggest that the violence and lawlessness in this country go far deeper than stolen TV sets and burned-out storefronts. The Bible sees the oppression of the poor rooted in the hardness of heart of the rich and comfortable, and in the institutions and structures they set up to further their own wealth and power. Again and again in the Scriptures, the exploitation and suffering of the poor is directly attributed to the substitution of the worship of mammon for true worship of God.[2]

The poorest of the poor have been shut out of American life. Increasingly, they are behaving as outsiders who have no stake in the society. America has been divided into two separate and drastically unequal worlds: the affluent majority and the impoverished class, the "we" and the "they." And the poor are even more desperate in other parts of the world.

I will never forget the first time I visited a shantytown on the outskirts of a Third World city. People lived in makeshift shelters made of discarded metal and wood scraps, tar paper, and cardboard. There was no sanitation, no running water, no electricity, and little food. Women hauled unclean water from two miles away. I didn't have to ask what kind of education the children were getting. Disease, illiteracy, high infant mortality, and, of course, hunger were the daily experience of the people. In the worst sections of Detroit, Chicago, the District of Columbia, or the South Bronx, I have never seen such human misery. I was told later that the government had come in two weeks after our visit and leveled the shantytown. The people were given fifteen minutes to pack up and clear out before officials set fire to the whole place and burned it to the ground. With no place to go, some would try to return to their former land in the rural areas, land that had been taken over by foreign corporations. Most, however, would simply put up another shantytown somewhere else. And the cycle would repeat.

The poor in the so-called underdeveloped countries of the southern hemisphere are the marginalized millions of our global economic system. They are at the bottom of the world social order and everybody's priority list; that is, if they make the list at all. They are the exploited urban workers whose cheap labor makes foreign investment so profitable to Western corporations. They are the rural people who have lost their land to national oligarchs and Western multinationals; many have been reduced to migrant labor for U.S. agribusiness. They work the mines of Africa, the sweatshops of Asia, and the plantations of Central and South America. They have become expendable. The lives of the world's poor move, back and forth, between the two poles of exploitation and abandonment.

Despite all the rhetoric about U.S. generosity and liberal foreign aid, the flow of world resources is overwhelmingly one-way, from the poor countries to the rich countries. Multinational corporations invest heavily in the poor countries for one simple reason: It is enormously profitable.[3] In addition to the cheap labor, the raw materials and energy resources required by the industrialized nations are often located in the poor countries. Our "right" to those resources has been a clear assumption of U.S. foreign policy, an assumption that has been backed by the use or threatened use of force. In addition, the United States usually claims the primary right to decide what to pay other countries for their resources, even if the price is far below Western market values. American consumers became so accustomed to this arrangement that they angrily cried foul when the OPEC countries claimed that primary right for themselves in the 1970s.

The pattern of land control has been a particularly cruel injustice in most poor nations. Some of the most tragic sights in the world today are pictures of starving people in countries that could be agriculturally self-sufficient but never will be while their land is used instead to grow cash crops for export to affluent nations. In the present international economic order, coffee, sugar, pineapples, tobacco, and bananas for rich people are a higher priority than bread and rice for the poor. Like housing in inner-city Washington, wherever life's necessities are made into profitable commodities, the poor will be denied what they need to survive. In the most impoverished places on the earth, the poor are simply dying. And they are dying young. This is the starkest reality of our historical situation.

Throughout history, the rich have had a difficult time seeing that their prosperity is based on other people's poverty. Basil, the fourth-century bishop of Caesarea, expressed his frustration on this point, saying to the rich, "How can I make you realize the misery of the poor? How can I make you understand that your wealth comes from their weeping?"[4] We don't make the connection either. We don't understand that we have much more than we need because the poor have much less than they need. We consume the resources of the earth far out of proportion to our numbers, while others go hungry and die for lack of life's basic necessities. In other words, our standard of living is rooted in injustice. Our hope is others' despair; our good life perpetuates their misery.

The question to be asked is not What should we give to the poor? but When will we stop taking from the poor? The poor are not our problem; we are their problem. The idea that there is enough for everyone to live at our standard of living, or that we are rich because of hard work and God's favor, or that poverty is due to the failures of the poor—all these are cruel myths devised by a system seeking to justify its theft from the poor. Billions of people are in imminent danger of simply being abandoned by an increasingly centralized global economic order that revolves around an affluent minority. That economic order has even pitted itself against the natural order. Charged to be stewards, we have instead become exploiters. Rather than treating the bounty of the earth as a gift for all of God's children, we have wasted its resources to profit the few.

The imperialism of today differs from the old colonialism in that empire is no longer based on the occupation of territory but on the control of resources. The world economy is dominated by the wealthy nations and arranged to guarantee them the largest share of the earth's goods and benefits. The United States still leads an international economic system in which twenty percent of the people control 86 percent of the wealth.[5] The U.S. definition of the "free world" is any place where U.S. interests can operate freely; many of those places have been ruled by regimes friendly to us but brutal to their own people.

Why is the United States the richest nation on earth? The answer is due neither to Yankee ingenuity nor to God's special blessing. It is starkly material. As our nation grew, resources were always cheap. We built our country on land taken from Native Americans and the labor of black slaves and, later, ethnic immigrants. When our westward movement reached the boundaries of the Pacific, we sought new territories for commercial conquest overseas. We constructed a foreign policy of economic penetration backed by military power. If financial

and political pressure failed, we could always send in the
Marines.

The U.S. empire reached its peak during the years after the
Second World War, from 1945 until the mid–1970s. During
that period the United States intervened in the affairs of other
nations—economically, politically, and militarily—more than
any nation on earth, including the Soviet Union. To turn
from our need for control in the world is the first step toward
national healing and conversion.

> O my people, your leaders mislead you, and confuse the
> course of your paths. The Lord has taken his place to
> contend, he stands to judge his people. The Lord enters
> into judgment with the elders and princes of his people:
> "It is you who have devoured the vineyard, the spoil of
> the poor is in your houses. What do you mean by
> crushing my people, by grinding the face of the poor?"
> says the Lord God of hosts. (Isa. 3:12b–15)

Rich Americans have reason to fear such judgment. But
they ignore it in favor of a different kind of fear, which increasingly plagues them: the fear of losing control in a changing world. Resources are no longer cheap. International
events are not so easily influenced as before. Insurgency and
terrorism is on the rise in many nations and is directed right
at the wealthy nations. At home, wages for many working
families have stagnated, while the costs of health care and
housing skyrocket. The American public, so highly trained in
the ways of consumption, is frightened and insecure. Our
standard of living, once assumed not only to last forever but to
increase forever, is threatened. Parents no longer trust that

their children will have a better life than they have had. The slogans and symbols by which we have justified ourselves and our place in the world seem shaky now. Most Americans have no real idea of a different way to live.

The result of prosperity based on injustice is anxiety. Gloom, cynicism, despair, and hedonism are all fruits of the fundamental anxiety that characterizes the cultures of the wealthy nations. The spiritual crisis of the rich countries directly corresponds to the economic crisis of the poor countries. The rich hunger in spirit while the poor hunger for bread. Our spiritual malaise is the consequence of affluence in the face of deprivation. Conversion in our time is to liberate the poor and to make the blind see. The poor need justice, and the rich need restored sight.

The anxiety-ridden and the poverty-stricken are the two major segments of humanity in the world today. To recognize this fact is to acknowledge the truth of the oppressor-oppressed division but to go beyond class consciousness to spiritual consciousness. The spiritual state of the affluent is thus clarified as well as their class position. The dictionary definitions of two common words provide valuable insight here: *Possessed*—influenced or controlled by something (as an evil spirit or passion); urgently desirous to have something. *Dispossessed*—those ejected, ousted, or deprived of possession (as of land).

The words describe the two major groups of people in the world today. The affluent are literally possessed by their possessions. Money and the things it can buy stalks the rich countries like a demon. Mammon offers comforts and pleasures to delight the flesh but demands the soul in return. The attachment of Americans to their standard of living has become an addiction. We can't stop shopping, eating, consuming. We laugh at how foolish and frivolous advertising has become and then, as if led by a Pied Piper, follow the sweet music of consumption to our next purchase. A successful life leads not to love, wisdom, and maturity; progress and success in our society is instead based on adding more to one's pile of possessions. Our natural course is toward a better job, bigger house, and richer lifestyle. Any variation in that plan of life is universally viewed with bewilderment if not open suspicion. The values of self-discipline and self-sacrifice have become so countercultural that they have virtually disappeared. Living within limits has no meaning in a society where any idea of limitations is held up to ridicule.

That which commands our time, energy, and thoughts is what we really worship. The things we usually think about, worry over, and plan for are the things we value most. Listen to the conversation of most middle-class Americans. A very large part of it revolves around consumption: what to buy, what was just bought, what products are preferred, where to eat, what to eat, the price of the neighbors' house, what's on sale this week, our clothes or someone else's, the best car on the market this year, where to spend a vacation. . . . To help us in these big decisions of life, there is the constant barrage of commercials that sound increasingly theological: "Datsun Saves," "Buick, Something to Believe In," "Kmart Is Your Saving Place," "Keep That Great GM Feeling," "The Good News of Home Heating," "GE: We bring good things to life."

Material goods have become substitutes for faith. It's not that people literally place their cars on the altar; rather, it is the function of these goods in a consumer society. They function as idols, even though most affluent U.S. Christians, like rich Christians throughout history, would deny it.

But I've never made an idol . . . nor set up an altar nor sacrificed sheep nor poured libations of wine; no, I come to church, I lift up my hands in prayer to the only-begotten Son of God; I partake of the mysteries, I communicate in prayer and in all other duties of a Christian. How then . . . can I be a worshipper of idols.[6]

That's how John Chrysostom, the fourth-century patriarch of Constantinople, said the rich responded to the charge of idolatry. He told them, "You pretend to be serving God, but in reality you have submitted yourself to the hard and galling yoke of ruthless greed."[7] Today, as then, material goods and wealth have assumed the place of ultimate concern, that is, the place of worship.

The tragic irony is that most Americans feel that they are "just getting by." No matter how much they have, they continue to protest, "We're barely keeping up with the bills." Most are up to their necks in credit card debt. Even if they wanted to get out, it would take years. They are, indeed, trapped. Or, in more biblical language, they are in bondage. The people who have more money and goods than any people in the history of the world spend most of their time worrying about not having enough. We have come to hold all the values of the consumer system without recognizing our subservience to it—the most perfect form of slavery. Cyprian, the third-century bishop of Carthage, described the rich this way:

Their property held them in chains . . . chains which shackled their courage and choked their faith and hampered their judgment and throttled their soul. . . . If they stored up their treasure in heaven, they would not now have an enemy and a thief within their own household. . . . They think of themselves as owners, whereas it is they rather who are owned: enslaved as they are to their own property, they are not the masters of their money but its slaves.[8]

Perhaps all the talk of freedom in this country is just a desperate attempt to convince ourselves that we still have some.

The affluent are the possessed. We have already spoken of the dispossessed, those who are preoccupied not with tomorrow's purchase but with today's survival. The worlds of the possessed and the dispossessed are separated by more than miles. Most of the rest of this chapter is concerned with bridging that gap. We will discuss how the rich could be, and biblically should be, converted to the poor. We begin with a fact almost obliterated by the world's present economic realities: the spiritual fact that all of us, rich and poor and in-between, are one humanity. The absolute value of every human being must be regained if we are to have any future. Our solidarity with one another has been dramatically shattered. Our fellowship with one another must be restored.

THE CONTACT OF THE affluent with the poor today is primarily through two means, television and statistics. We hear the stark statistics of human suffering and we watch starving children in living color. But what do those numbers mean to us, and how real are the young lives we glimpse for a moment in a news documentary? A very wise old man once told me the difference between concern and compassion: "Being concerned is seeing something awful happening to somebody and feeling, 'Hey, that's really too bad.' Having compassion," he said, "is seeing the same thing and saying, 'I just can't let that happen to my brother.'" In other words, concern comes from a recognition of a problem. Compassion comes out of a feeling of relationship.

That is precisely what the affluent lack with respect to the poor—any real feeling of relationship. We have no relationship with the poor because we have no proximity to the poor. Structurally and personally we isolate ourselves by isolating "them." In the United States, suburbs are not just a place to live surrounded by trees and green grass; they are also a way to get away from poor people. Great modern superhighways carry us over and around the sights, sounds, and smells of poverty. In the natural course of their days, the affluent increasingly pass by the human faces of poor brothers and sisters.

The pictures and statistics of the television documentaries may inform, shock, repulse, or immobilize us. They may even be enough to make us feel guilty; but they don't change our hearts, which, in the biblical language, means to change our lives. Identification with the poor is, after all, impossible without some concrete relationship to the poor. "But if anyone has the world's goods *and sees his brother in need*, yet closes his heart against him, how does God's love abide in him?" (1 John 3:17; italics mine). What if the course of our lives is so arranged that we never see the brother or sister in need? Or, in the following question of John Chrysostom, do we injure ourselves less by neglecting the poor whom we've made relatively invisible? "Don't you realize that, as the poor man withdraws silently, sighing and in tears, you actually thrust a sword into yourself, that it is you who received the more serious wound?"[9]

Most people who have begun to make radical changes in their economic lifestyle have had some personal experience of the woundedness of the poor. Proximity to poor people is crucial to our capacity for compassion. Only through proximity do we begin to see, touch, and feel the experience of poverty. When affluent people find genuine friendships among the poor, some revolutionary changes in consciousness can begin to take place.

I remember a friend named Butch. We worked together in Detroit. We were the same age, but I was white and he was black. All my money, I saved for college. All of his went to support his wife, mother, and younger brothers and sisters. He was smart and knew a lot more about the world than I did. Yet he could barely write down the directions to his house the first time I visited. I met his mother. Like my mother, she cared about her son's safety, and she feared that his militance would get him in trouble. I'll never forget what she said about the police. My mother always told us to look for a policeman if we ever got lost. Butch's mother told her children to hide from the police if they were lost. Everyone in Butch's family told me personal stories of police racism and abuse. It jarred me. Proximity opens our eyes and changes our hearts.

I always get stuck when someone asks me how they can identify with the poor without coming to where the poor live. Now, I don't believe the answer is for everyone to move into the inner city. However, I know of no way to have our lives really affected by the poor other than through relationship to the poor. How else can one begin to listen, to learn, to understand? That, by the way, is the *only* posture in which the affluent can begin to approach poor people. Not to tell, nor teach, nor even help, but to put themselves in a position to *listen* and to be changed.

That has been the experience of Sojourners community. After many years in the inner city, we had fewer and fewer illusions about our identification with the poor or about how much help we could bring. The longer we were in our D.C. neighborhood, the more aware we were of how different our lives were from the lives of our neighbors. We lived simply, but all pretensions of "being poor" vanished. The differences between voluntary and forced poverty are immense. But one thing is true: the poor were ever before us. We learned that we needed that. By living where and how we did, their existence could not be ignored. The key for the affluent is to be in a situation where they can no longer ignore the existence of poor people. Society's denial of the existence of the poor is at the root of their oppression. If we put ourselves in daily relationship to poor people we will understand more than oppression; we will begin to understand the very sufferings of Christ. It was always hard for our community in Washington to do that. But even when we wanted to ignore the pain and violence of our neighborhood, our presence there made it impossible for us to do so. The daily realities of inner-city life were quick to intrude and interrupt our lives. In my better moments, I saw it as a holy intrusion; at other times, it was more like a pain in the neck. Sometimes I would rather find Jesus in some other form than in the kids on the street, the evicted family that again needs a place to stay, the troubled people with innumerable problems that take so much time. But Jesus is there, waiting for me to recognize him and invite him in. One of our community members wrote a song about that. Here are some of the words:

Is there room in the inn for a child's life to begin?
Is there room in this world for a suffering servant King?
Is there room in our life to give you all honor and
    praise?

Well, come in, Jesus Child. We want to make you
   some room.

Is there room in your home for the outcast family?
Is there room in your heart for the lonely one?
Is there room at your table for the hungry one?
Well, come in, Jesus Child. We want to make you
   some room.

Is there room in this city for the lowly and the poor?
Is there room in this city for the homeless and their
   friends?
Is there room in this life for the broken little ones?
Well, come in, Jesus Child. We want to make you
   some room.*

Putting ourselves in proximity to oppressed people begins
to open us up to understanding and compassion. Our hardness
of heart cannot be maintained for long in the midst of such
obvious human suffering. Only when our eyes are opened and
our hearts softened are we really ready to work with and for
the poor in seeking to change the wretched circumstances of
their lives. God identifies with the poor, not because they are
more noble but because they are more vulnerable. The rich
need fewer noble intentions and more open vulnerability.
Proximity will help to remove our hearts of stone and replace
them with hearts of flesh. This is radical surgery for the affluent, surgery finally that only God can do. In other words, we

---

* By David McKeithen. Copyright © 1979 by Sojourners Fellowship.

need to be open to conversion. The Scriptures make our need for conversion on this issue utterly clear.

I am an evangelical Christian. The word *evangelical* is a good one. It has its origins in the root word *evangel*, which means "good news." In fact, the word translated in the New Testament as "evangelist" is the noun from a verb that means "to announce the good news." To be an evangelical Christian, then, means to identify oneself with the good news that Jesus preached, namely, the gospel of the kingdom of God. Christ's inaugural sermon in the little town of Nazareth made clear how, and why, and to whom his message was such good news:

> The Spirit of the Lord is upon me, because he has anointed me to preach good news to the poor. He has sent me to proclaim release to the captives and recovering of sight to the blind, to set at liberty those who are oppressed, to proclaim the acceptable year of the Lord. (Luke 4:18–19)

To the Jewish masses under the yoke of Roman domination, the message was good news indeed. The ruling religious and political authorities, however, found this evangel very bad news and set themselves against Jesus right from the beginning.

Today, the greatest number of the world's people—the poor and oppressed—also would find the evangel of Jesus to be the best news they had heard in a long time. But is that the message heard today from our evangelical preachers? Do the affluent millions who comprise the burgeoning American evangelical movement find their identity in the promise of salvation, freedom, healing, and liberation proclaimed by Jesus at the outset of his ministry? Does the word *evangelical* conjure up the vision of a gospel that turns the social order upside down? Listening to modern evangelical proclamations leaves one with the distinct impression that the content of the message has been changed. The image of American evangelicalism that comes out of many pulpits and goes out over the air waves is a religion for those at the top, not those at the bottom of the world system. It bears almost no resemblance to the original evangel.

It has not always been so. Evangelical movements in England and the United States have led struggles for the abolition of slavery, for economic justice, and for women's rights. Eighteenth-century English preachers and nineteenth-century American evangelists deliberately linked revivalism to social change and proclaimed a gospel that was indeed good news to the poor, the captives, and the oppressed.[10]

However, in what sociologist David Moberg has called "the great reversal," twentieth-century evangelicalism in the United States came to identify thoroughly with the mainstream values of wealth and power.[11] As the country became rich and fat, so did its evangelicals, who soon replaced the good news of Christ's kingdom with a personal piety that comfortably supported the status quo. Evangelicals in most of the twentieth century were not known as friends of the poor. Rather, evangelicals were known to have a decided preference for the successful and prosperous who saw their wealth as a sign of God's favor. Ironically, a movement that once fought to free slaves, support industrial workers, and liberate women came to have a reputation for accommodating racism, favoring business over labor, and resisting equal rights for women. In our nation's ghettoes, barrios, and unions, evangelicals were generally not regarded as allies.

The good news of the twenty-first century is that things are beginning to change. A new generation of evangelicals is recognizing that, as Rick Warren recently wrote: "global poverty is an issue that rises far above mere politics. It is a moral issue . . . a compassion issue . . . and because Jesus commanded us to help the poor, it is an *obedience* issue! He told us to do all we can to alleviate the pain of our brothers and sisters: *'Inasmuch as you did it to one of the least of these my brethren, you did it to me' (Matthew 25:40, NKJV)."*

The "acceptable year of the Lord" Jesus came to proclaim was, according to many New Testament scholars, a reference to the Jubilee year of the Old Testament, which provided for a periodic redistribution of land and wealth along with the freeing of slaves.[12] The Jubilee was a corrective measure aimed at our sinful human tendency toward the accumulation of wealth at the expense of the poor. Evangelicals today, however, have not been the ones calling for economic redistribution. Instead, they have tended to favor tax breaks for the middle class and for the big corporations. Many usually support increased military spending and budget balancing by cutting the amount of public resources available for the poor. Simplicity, stewardship, and redistribution are all biblical values no longer associated with the evangelical message. The evangelical nationalists exalt the nation at a time when the United States needs to be humbled. They extol the virtue of wealth and power when most of the world is poor and powerless. They call for unrestrained economic growth in a world where resources are running out and much of God's creation is ravaged by industrial exploitation.

We must recover the evangel. The public image of evangelicalism in this country is a distortion of the best of that tradition. As we recover that tradition we will recover our bearings in the Bible. In the early days of our community, we did extensive biblical studies on the question of wealth and poverty. We studied a number of good books that were published on the subject in response to growing revelations of widespread world hunger and the great contradictions of American affluence.[13] Our investigations into the biblical material produced some startling conclusions.[14]

First, we were impressed with the sheer bulk of biblical teaching on the subject of wealth and poverty. It is one of the most central themes in all of Scripture. It pervades the Old Testament and, some have even suggested, is the second most common topic found there, the first being idolatry. Most often, the two are directly related. In the New Testament, we found more than five hundred verses of direct teaching on the subject. That is one out of every sixteen verses, and that ratio does not include indirect teaching drawn from New Testament doctrines and the actions of Jesus or the apostles. Jesus talked more about wealth and poverty than almost any other subject, including heaven and hell, sexual morality, the law, or violence. One out of every ten verses in the Synoptic Gospels is about the rich and the poor; in Luke, the ratio is one out of seven. James treats the subject in one out of every five verses in his epistle. Thus, the subject of money, possessions, and the poor is hardly a casual concern or passing interest to the biblical writers. The Bible is literally filled with it. In the midst of our study, we began to wonder why the topic had been so completely ignored in all of our church upbringings.

Not only is the Bible strong in its emphasis; the Scriptures are stunning in their clarity on this issue. Wealth is seen, at best, as a great spiritual danger and, most often, as an absolute hindrance to trust in God. The rich are continually held responsible for the sufferings of the poor, while God is portrayed as the deliverer of the oppressed. The God of the Bible has taken sides on this matter and has emphatically chosen the side of the poor. Sharing with the poor is not regarded as an option but as the normal consequence of faith in God.

Biblical economics begins with the affirmation that the earth is the Lord's. Its care is given to humankind in sacred trust; its sustenance is to be shared by all God's children. The biblical doctrine of stewardship renders a clear judgment against any economic system based on ever-expanding growth, profit, and exploitation of the earth. The Old Testament sees poverty as neither accidental nor natural but rooted in injustice. The prophets railed against the rich for their oppression of the poor.

> Hear this word, you cows of Bashan, who are in the mountains of Samaria, who oppress the poor, who crush the needy, who say to their husbands, 'Bring, that we may drink!' The Lord has sworn by his holiness that, behold, the days are coming upon you, when they shall take you away with hooks, even the last of you with fish-hooks. (Amos 4:1–2)

Yahweh demanded justice and righteousness and declared that nations would be judged by how they treated their poor. The early Hebrew codes built in provisions for periodic redistribution to counter the concentration of wealth and to ensure equity and justice. Right relationship to the Lord required the setting straight of all economic and social relationships.

Jesus is God made poor. His coming was prophesied to bring social revolution, and his kingdom would turn things upside down: The mighty would be brought low, the rich sent away empty, the poor exalted, the hungry satisfied (Luke 1:52–53). Jesus identified himself with the weak, the outcast, the downtrodden. His kingdom undermines all economic systems that reward the rich and punish the poor.

The early Christians shared their goods with one another and with the poor. The Jubilee redistribution was fulfilled among them, no longer just at periodic intervals, but as a way of life. The apostles taught that one could not profess love for God while ignoring the needs of hungry neighbors. The problem, according to the Old Testament, is the *accumulation* of wealth and the oppression of the poor. The New Testament carries the theme further, regarding the very *possession* of wealth as a fundamental spiritual problem; for wealth distorts people's priorities, makes them insensitive to others, and seriously obstructs their relationship to God. Simply having undistributed wealth, in the midst of human need, is unequivocally condemned in the New Testament. And what the Bible condemned then, it condemns now. Everything the Bible says about the sins and obligations of the rich applies directly to us as affluent Americans.

Jesus knew the importance of money and possessions in human society, that questions regarding money go right to the heart of each individual, and he treated the issue as a deeply spiritual matter. He maintained that our true values and priorities are shown in our relationship to the poor and to material goods. Our real commitments are revealed in our economic choices. His words on the subject are uncomfortably clear.

Woe to you that are rich, for you have received your consolation. Woe to you that are full now, for you shall hunger. (Luke 6:24–25)

It is easier for a camel to go through the eye of a needle than for a rich man to enter the kingdom of God. (Matt. 19:24)

So therefore, whoever of you does not renounce all that he has cannot be my disciple. (Luke 14:33)

Other seed fell among the thorns and the thorns grew up and choked it, and it yielded no grain. . . . [T]hey are those who hear the word, but the cares of the world, and the delight in riches, and the desire for other things, enter in and choke the word, and it proves unfruitful. (Mark 4:7, 18b–19)

Take heed, and beware of all covetousness; for a man's life does not consist in the abundance of his possessions. (Luke 12:15)

In his Sermon on the Mount, Jesus speaks extensively about wealth, worship, and trust.

Do not lay up for yourselves treasures on earth, where moth and rust consume and where thieves break in and steal, but lay up for yourselves treasures in heaven, where neither moth nor rust consumes and where thieves do not break in and steal. For where your treasure is, there will your heart be also. (Matt. 6:19–21)

That's not what we hear in the churches. Today, we hear, "Where your heart is, that's where your treasure will be." We think: The problem is not so much the nature or amount of our possessions, it's our attitude toward them. We believe: As long as we are inwardly free, as long as we are *willing* to give it all up, the possessions are not a problem. We rationalize: As long as our hearts are in the right place, it doesn't matter how much we spend or how much we accumulate. But Jesus doesn't seem to agree with our self-justifications. The problem of possessions is not nullified by a "proper attitude" toward them. Jesus says, "Where your wealth is, that's where your heart is."

Jesus continues: "No one can serve two masters; for either he will hate the one and love the other, or he will be devoted to the one and despise the other. You cannot serve God and mammon" (Matt. 6:24). Here Jesus is not giving advice. He is not saying, "You should not try to serve both God and money." He is not saying, "It's not a good idea; I'd rather you didn't." Jesus is saying that you *cannot* do it. It will not work. You cannot serve the two masters at the same time; they make rival claims on our lives, time, energy, and resources.

Jesus goes on to teach about anxiety and security.

> Therefore I tell you, do not be anxious about your life, what you shall eat or what you shall drink, nor about your body, what you shall put on. Is not life more than food, and the body more than clothing? Look at the birds of the air: they neither sow nor reap nor gather into barns, and yet your heavenly Father feeds them. Are you not of more value than they? And which of you by being anxious can add one cubit to his span of life? And why are you anxious about clothing? Consider the lilies of the field, how they grow; they neither toil nor spin; yet I tell you, even Solomon in all his glory was not arrayed like one of these. But if God so clothes the grass of the field, which today is alive and tomorrow is thrown into the oven, will he not much more clothe you, O men of little faith? Therefore do not be anxious, saying, 'What shall we eat?' or 'What shall we drink?' or 'What shall we wear?' For the Gentiles seek all these

things; and your heavenly Father knows that you need them all. But seek first his kingdom and his righteousness, and all these things shall be yours as well. (Matt. 6:25–33)

We are mistaken if we think that this passage merely talks about economics or about social justice. The passage is about whether or not we trust God. Do we finally know ourselves to be God's children at such deep levels that we can trust God for our material needs? This is part of the cost of discipleship. That cost is to trust in God. Following the Master means sharing his life, foregoing worldly securities for the treasures of the kingdom.

To trust fully in God requires not only that we break our attachment to possessions, but also that we identify ourselves with the poor and the afflicted in their distress. To do so is not just to follow the example of Jesus; it is to minister to Christ himself.

When the Son of man comes in his glory, and all the angels with him, then he will sit on his glorious throne. Before him will be gathered all the nations, and he will separate them one from another as a shepherd separates the sheep from the goats, and he will place the sheep at his right hand, but the goats at the left. Then the King will say to those at his right hand, 'Come, O blessed of my Father, inherit the kingdom prepared for you from the foundation of the world; for I was hungry and you gave me food, I was thirsty and you gave me drink, I was a stranger and you welcomed me, I was naked and you clothed me, I was sick and you visited me, I was in prison and you came to me.' Then the righteous will answer him, 'Lord, when did we see thee hungry and feed thee, or thirsty and give thee drink? And when did we see thee a stranger and welcome thee, or naked and clothe thee? And when did we see thee sick or in prison and visit thee?' And the King will answer them, 'Truly, I say to you, as you did it to one of the least of these my brethren, you did it to me.' Then he will say to those at his left hand, 'Depart from me, you cursed, into the eternal fire prepared for the devil and his angels; for I was hungry and you gave me no food, I was thirsty and you gave me no drink, I was a stranger and you did not welcome me, naked and you did not clothe me, sick and in prison and you did not visit me.' Then they also will answer, 'Lord, when did we see thee hungry or thirsty or a stranger or naked or sick or in prison, and did not minister to thee?' Then he will answer them, 'Truly, I say to you, as you did it not to one of the least of these, you did it not to me.' And they will go away into eternal punishment, but the righteous into eternal life. (Matt. 25:31–46)

The passage is a picture of the risen, glorified Christ, sitting in judgment of those who call him Lord. He is the Son of God who is found among the suffering and the poor. There is his natural habitation, yet it is a place where the people never thought to look for him. Jesus has so cast his lot with the least of the earth that to serve them is to serve him; to ignore, neglect, and abuse them is to ignore, neglect, and abuse him. He stands among them, and he asks us not just how much we love the poor; Jesus stands among the poor and asks all those who would call his name, "How much do you love me?"

The Matthew 25 passage is a statement of love for the poor that is more radical than anything in secular literature. God loves all people in all circumstances and certainly loves rich people. But God shows a special concern and compassion for the poor. God loves the rich and the poor as persons, but the Bible speaks against the rich as a class of people and takes the side of the poor.

The Bible often refers to the oppressed, the alien, the stranger, the orphan, and the widow. These are the defenseless people, the powerless, the disenfranchised, the voiceless ones at the bottom of the social structure. But by his relationship with the poor, Jesus establishes their value. So must the church. The Christian point of view must be that of those at the bottom. Their rights and needs should always be the most determinative element of the church's social stance.

If Jesus identified himself with the poor, what does that mean for us? What is the place of the poor in the priorities of our lives? Jesus Christ is known for his position on that question. If we are his people, we must be known for the same thing. We simply join him where he has already cast his lot. Our relationship to the poor is not the only issue here. Our spiritual well-being and our relationship to the Lord are at stake.

Perhaps the most famous and striking of Jesus's teachings on the subject of wealth took place when he met the rich young ruler.

And as he was setting out on his journey, a man ran up and knelt before him, and asked him, "Good Teacher, what must I do to inherit eternal life?" And Jesus said to him, "Why do you call me good? No one is good but God alone. You know the commandments: 'Do not kill,

Do not commit adultery, Do not steal, Do not bear false witness, Do not defraud, Honor your father and mother.' " And he said to him, "Teacher, all these I have observed from my youth." And Jesus looking upon him loved him, and said to him, "You lack one thing; go sell what you have, and give to the poor, and you will have treasure in heaven; and come, follow me." At that saying his countenance fell, and he went away sorrowful; for he had great possessions. (Mark 10:17–22)

This story is one of my great favorites. There was a time when I would preach from this passage in a wealthy congregation and really "lay them out." By the end of the sermon, most people would be feeling very guilty and low about themselves and their lives—they had been convicted. And I would feel quite prophetic. I could walk away from these congregations time and again, knowing I was still on the cutting edge. A few years later, however, two elements in this passage that I never saw before became clear to me. The first is that Jesus looked at the young man and loved him. I believe that young man knew he had been loved that day. He turned away, not just from one who brought him the meaning of the kingdom in relation to his possessions; he also turned away from one who clearly did love him.

There is another part of the passage that I had somehow missed. I always stopped at the point where Jesus told the young ruler to sell all that he had and give it to the poor. I simply told the wealthy Christians that this was what they too had to do. I was saying, in effect, "Sell all that you have, give it to the poor, and drop me a card to let me know how everything comes out." For people who are trapped, who are under the control of their possessions, who are up to their necks in debt, who have more than anyone ought to have yet are still fearful of not having enough—for these people, my words of "hand it all over" were insensitive and missed the full message. For Jesus did not stop there. In addition to "go, sell what you have, and give to the poor," he said, "and come, follow me." Whenever Jesus says come and follow him, he is saying: I'd like to invite you to join my community. I'd like you to pick up your life, to pack up your bags and come share your life with us. Come join us, come experience the new kind of security that we have found by trusting God together. You'll find people very much like yourselves, people with the same fears, problems, and history, but people who want to learn a new way of living together, friends who will take you by the hand, and together, will be led with you to a new place. Come and live among us and learn what the new order is all about.

If we care about breaking the stranglehold that our economic system has on the church's life, we will need to create communities that invite people in, places where repentance is more than a word, where a concrete and visible alternative begins to be demonstrated. This is the meaning of conversion's call to repentance as Jesus offers it. Repentance is never divorced from the creation of a new people whose life together is to be the firstfruit, the pilot project, the seed of the new order that God intends for the whole of creation. If our renewal does not generate the power to break free from the hold that mammon has on the church's life in this country, it is not worth talking about. But if we can experience the Spirit in the way the early Christians did, if we can trust God for our needs and stand with Christ among the poor, then the power of our conversion will be plain for all to see.

There are certain passages of the Bible that Christians have exerted considerable effort to avoid. If a most-avoided list were compiled, the passages in Acts about the economic sharing of the early church would rank high. For years the church has done theological acrobatics to avoid the implications of what the first Christians did when the Spirit invaded their lives.

> And all who believed were together and had all things in common; and they sold their possessions and goods and distributed them to all, as any had need. And day-by-day, attending the temple together and breaking bread in their homes, they partook of food with glad and generous hearts, praising God and having favor with all the people. (Acts 2:44–47a)

> Now the company of those who believed were of one heart and soul, and no one said that any of the things which he possessed was his own, but they had everything in common. And with great power the apostles gave their testimony to the resurrection of the Lord Jesus, and great grace was upon them all. There was not a needy person among them, for as many as were possessors of lands or houses sold them, and brought the proceeds of what was sold and laid it at the apostles' feet; and distribution was made to each as any had need. (Acts 4:32–35)

Theologians and preachers offer various explanations of these two passages. Some say that the early church members, because they expected Christ to come back in their lifetimes, developed an interim ethic until his return. Their economic sharing, therefore, was temporary and not normative. They were mistaken in their apocalyptic expectation. Jesus did not come back, so the more traditional economic pattern had to be re-instituted.

Some people say that the sharing of the early Christians was in response to an economic crisis, a state of emergency, and therefore does not apply to other times and situations. Others say this sharing was simply an early, experimental model for the church's life, and that it failed. Not being a people who like failure, we can safely disregard it and keep our economic lives private. But it didn't simply fail. Economic sharing continued to be taught and practiced by the church long after the Jerusalem experience. The *Didache,* a second-century manual of church discipline, said, "Share everything with your brother. Do not say, 'It is private property.' If you share what is everlasting, you should be that much more willing to share things which do not last."[15] Tertullian in the third century said, "We who share one mind and soul obviously have no misgivings about community in property."[16] These are but two examples of the outspokenness of the early church fathers on the calling of Christians to share their earthly goods, both within the believing community and with all who are in need. As leaders of the church in the first few centuries of its life, the church fathers consistently adopted a radical position on economic matters. Their position reflected their pastoral concern for the spiritual well-being of Christians and their compassion for the poor. The spirit and example of Jesus fills their writings. Their testimony contradicts the commonly held opinion that Jesus's teachings on this subject were only an ideal that was quickly moderated and replaced with more realistic economic values.[17]

Others, however, want to find in the Acts passages a new rule for the church, a new law that states that the road to renewal is paved by the common purse. Their idea is that if you just take all your money and throw it in the pot, you will be renewed. That is neither true nor the point of the passages.

Despite all the attempts to rationalize the meaning of the verses quoted here, what they mean is not that complicated: they are simply descriptions of what happened when the Holy Spirit invaded the lives of the early Christians. The coming of the Spirit among them shattered the old and normal economic assumptions and created an entirely different economic order. The Spirit established a new way of thinking and a new way of living that affected their relationship to money and goods. It created a new economy among the Christians, a common life in which economics was no longer a private matter; economics was a matter of fellowship, in fact, a central matter of fellowship. The early Christians could not have conceived of a way to share their lives spiritually without sharing their lives economically. The sharing recorded in Acts 2 and 4 was the attempt by the early Christians to make practical their understanding of the mind of Christ on economic matters.

The relationship between the coming of the Spirit and the creation of the new economy is key. The Christians were making the teachings of Jesus real, and their decisions were empowered by the coming of the Spirit.

Later, Paul was taking a collection for the Jerusalem church, which was suffering from famine. In 2 Corinthians 8, he makes a vital connection between spiritual unity and economic equality. We are called to share, not just to be ready to share. Accumulating wealth while brothers and sisters are in poverty is evidence of sin in the church's life. Why should Christian congregations live such different economic lives in different places? Why should some Christians be poor while others have more than they need? Are we not family? Paul uses the word *equality* (verse 14). He wanted the Corinthians to understand that the unity of heart and mind God wanted them to have with their brothers and sisters in Jerusalem had to do with, among other things, economic equality between them. What does that say to us, rich Christians in a rich land, about a world of poverty in which many Christians and their churches are poor?

Our affluent way of life is not only crushing the poor economically. The consequences are also spiritual. In a world context of Christians who are poor, our way of life is a violation of Christian fellowship. Our wealth breaks the unity of the body of Christ. We, like the Corinthians, need to understand what Paul meant by equality.

The early Christians did not share their resources out of obligation, guilt, or in obedience to a new rule called "equality." They shared their goods out of a tremendous experience of joy and spontaneous offering. They had experienced the Holy Spirit in their midst, and their response was to share everything they had.

The Acts passages are important to many people in the church, but for very different reasons. Those committed to charismatic renewal find in this passage the centerpiece to all they are about; it describes the coming of the Spirit at Pentecost. There are others in the church concerned with justice and redistribution for whom this passage is also key. But each group manages to miss in the passage that which the other group finds. Why? Why is it that we continue to take the part and not the whole? Why do we act on false choices again and again? According to Acts, behaving differently about money is a visible consequence of the Spirit's presence—a very hard word for people who would rather stay in the upper room with the Spirit and never come down in the street to live out the new economy that the Spirit has created. Similarly, those who are economically aware and involved with the poor often don't recognize that what the Spirit has created has any relevance for them. It is the very presence of the Spirit, however, that breaks through our old assumptions, breaks down our self-interest, and makes the new economy possible. But for many, to talk of the Spirit is an embarrassment. It's awkward. It sounds too much like our pious pasts; it smacks of religious upbringings that we want to forget.

The lesson of Acts 2 and 4 is twofold. Any spiritual renewal that doesn't result in a new economy among the believers is incomplete, if not inauthentic. But also, the creation of the new economy in the church is not going to come about simply through good intentions, political analysis, educational programs, or social action projects. It will come about only when we begin to experience something of what the early believers did when the Spirit entered their lives and established a common life among them. The Spirit is the power of the new economy.

The gospel story of the widow's mite (Mark 12:41–44) makes a related point. The example of the new economy she demonstrated in her act of giving had little to do with her social analysis. It had to do with her relationship to God, which had transformed the economics of her life. The example of the poor widow's generosity illustrates an important New Testament principle about sacrificial giving: How much is given is less important than how much is left over after giving. Although that principle is completely reversed in most churches today, Jesus's high praise for this simple woman gives her a place of honor high above all wealthy church patrons.

God's siding with the poor and the oppressed is much clearer than most of the issues that have divided the church. The scriptural demand for economic and social justice is quick to be purged when the church accommodates to the established order. But, to know God is to love the poor and plead the cause of the oppressed. The proper vantage point for the people of God is to view institutions, systems, and societies from the perspective of their victims. The church visioned in the New Testament embraces the dispossessed and assumes the mantle of servanthood after the manner of God in Christ.

Today most of the world's people are poor—1.2 billion earn less than $1 per day; 852 million are hungry. Eleven million children under the age of 5 die every year, more than half of them from hunger-related causes.[18] That is devastating. In the midst of the devastation, God's son has taken up residence. John Chrysostom, noting the identification of Christ with the poor, said:

> You eat to excess; Christ eats not even what he needs. You eat a variety of cakes; he eats not even a piece of dried bread. You drink fine Thracian wine; but on him you have not bestowed so much as a cup of cold water. You lie on a soft and embroidered bed; but he is perishing in the cold.... You live in luxury on things that properly belong to him. Why, were you the guardian of a child and, having taken control of his estate, you neglected him in his extreme need, you would have ten thousand accusers and you would suffer the punishment set by law. At the moment, you have taken possession of the resources that belong to Christ and you consume them aimlessly. Don't you realize that you are going to be held accountable?[19]

The church in the United States is rich. Christians happily participate in our national lifestyle, which is consuming 40 percent of the earth's resources for a mere 4.5 percent of the world's people. That fact is a sin against Christ, if we believe the Bible. It sets our context, which is now our most important context—the one that shapes every other context of our lives. Resources are dwindling. The poor are becoming hungrier while the appetites of the rich show little sign of subsiding. The spirit of violence, actual and potential, escalates almost daily. All this now affects the church's life on every level—our evangelism, worship, pastoral care, and theology. When we see a church that has fallen into accumulation and affluence, a church that is being consumed by its own consumption, we are witnessing a church that has fallen into idolatry.

The church today is trying to serve both God and mammon, and the attempt has divided our loyalties. In a world where most people are poor, a rich church is living testimony of idol worship. The mere possession of such wealth is proof of serving money. We did not become affluent by sharing with the poor, but only through accumulation. The Bible calls that slavery, a relationship of bondage. Our accumulation has put us in servitude to mammon. A wealthy church cannot testify to dependence on God. God's people have forgotten who they are, forgotten to whom they belong, forgotten what it means to worship the Lord.

The decades leading up to the American Civil War saw a revival in this country. The "Great Awakening" was marked by many conversions and a mighty outpouring of evangelical faith. Central to the revival was an unswerving opposition to one of the greatest evils ever devised by human beings—the institution of slavery. As far as the revivalists were concerned, to turn to

Christ meant to turn from slavery. Opposition to slavery was seen as a fruit of conversion. The practice of making people into property was considered an abomination to God, a monstrous system that had to be stopped as a matter of faith.

Many converts, after committing their lives to Christ, enlisted in the abolitionist cause. The awakening, which reached its peak in the 1830s and 1840s, linked revivalism with abolitionism. The great preachers and leaders of the revival—Charles Finney, Jonathan Blanchard, Theodore Weld, the Tappan brothers, and the Grimke sisters—called for conversion that reconciled people not only to God but to one another.

Charles Grandison Finney was, first of all, an evangelist and has been called the father of modern evangelism. His preaching also "shook the nation and helped destroy slavery."[20] The failure of the churches to take a clear stand on the question of slavery was, in his opinion, one of the great "hindrances of revival." Donald W. Dayton describes Finney's stance:

> In his *Lectures on Revivals of Religion* Finney argued that "revivals are hindered when ministers and *churches take wrong ground in regard to any question involving human rights.*" He applied this particularly to slavery, insisting that "the church cannot turn away from this question." He argued that "the silence of Christians upon the subject is virtually saying that they do not consider slavery as a sin" and claimed that "it is vain for the churches to resist for fear of destruction, contention, and strife" or to "account it an act of *piety* to turn away the ear from hearing this cry of distress" from the "shackled and bleeding" slaves.[21]

When the church does not speak out on such an issue, said Finney, "she is perjured, and the Spirit of God departs from her."

Finney insisted that the church's response to slavery was a spiritual issue; the failure to embrace the slaves would undermine the spiritual life and integrity of the churches. "One of the reasons for the low estate of religion at the present time," he said, "is that many churches have taken the wrong side on the subject of slavery, have suffered prejudice to prevail over principle, and have feared to call this abomination by its true name." Finney regarded it impossible for the church "to take a neutral ground on this subject." He even went so far as to call for the use of church discipline against the sin of slaveholding. If the church "tolerates slaveholders in her communion SHE JUSTIFIES THE PRACTICE," he said. Finney himself took the strong position of refusing communion to slaveholders, saying, "where I *have authority*, I exclude slaveholders from the [Lord's Supper], and I will as long as I live."

Finney believed that the church's life would be destroyed from within by its support of slavery. His concern was for the spiritual vitality of the churches as well as for the freedom of the slaves. He was convinced that the church's position on the issue would, ultimately, affect the whole of its life, including its evangelism. At the center of his faith, Finney could not separate his call as an evangelist from his call as an abolitionist. In each case, his primary concern was the proclamation and the integrity of the gospel message.

Some historical issues stand out as particularly urgent among the church's other fundamental concerns. They are particularly threatening to human life and values and offensive to the righteousness of a loving God. These overarching moral questions intrude upon the routine of the church's life and plead for the compassion and the courage of God's people everywhere.

Slavery was such a question in the nineteenth century. Poverty is such a question today. Like slavery, our acceptance of poverty has brought us to a crisis of faith. Poverty is no more just a political issue than slavery was; it is a question that challenges everything we say about our faith in God and our allegiance to Jesus Christ. In other words, the massive nature of global poverty presents us with more than a test of our economy; it confronts us with a test of our conversion.

Conversion means to relinquish our wealth, for the sake of our relationship to God and for the sake of our fellowship with the poor. Justice requires an end to our accumulation. A new commitment to economic sharing and simplicity will both break our bondage to affluence and bring a vitality and integrity that most of our congregations have never experienced.

To share the control of our resources with fellow believers and to share what we have with the poor is part of what it means to be the church. Economic sharing is a way of life intended for all God's people, not just monastic orders, the clergy, or special ministry groups. No particular form or administrative procedure can ever be normative for all, but the biblical vision of fellowship will always include our economic lives, the sharing of our money and possessions with each other and with the poor.

The church will need to build a new economic framework for its own life if it hopes to impact this society. The severity of the current economic crisis makes the radical example of the early church even more crucial for us to understand and to apply today. Future wars will almost certainly be based on disagreements over world division of the economic pie.

The biblical material, the testimony of the early church, and our historical situation all lead us to the conclusion that a radical change in our relationship to money and possessions is a crucial test of our conversion.

And that test comes with a new altar call. For the first time the world has the knowledge, information, technology, and resources to end extreme poverty as we know it, but what is still lacking is the moral and political will to do so. I believe that generating such moral will is the vocation of the religious community. And, I believe that God is acting on the issue of poverty.

God is acting among us as religious leaders and faith communities, drawing us together as never before across theological and political boundaries in a moral, spiritual, and biblical convergence. In June 2005, an amazing procession of religious leaders from almost every major faith tradition in America joined in a service at the Washington National Cathedral. Anglican Archbishop Njongonkulu Ndungane of Cape Town, South Africa, noted the moral convergence of such a wide spectrum of American religious life and pronounced this a "kairos" moment—when regular time ("kronos") gives way to a spirit-filled moment in history and a new sense of time takes over.

God is acting in our culture. Artists and musicians are playing a critical role and a new generation of young people are committing themselves to this cause.

God is acting through new leadership in Africa where democracies are taking responsibility and acting in new ways to end corruption and create more transparent governance that makes effective aid more possible.

God is acting through the leaders of the world's wealthiest nations—the G8. The recent agreement to cancel forty billion dollars in debt for the world's eighteen poorest nations is a very important step. Our leaders should now finish the job by canceling all the debt for all impoverished nations.

Wealthy countries in Europe and the United States are beginning to do more in aid, but they must increase it still further. Many countries are now doubling aid. The goal for the United States of providing an additional 1 percent of its budget is a small price to pay for what is at stake—our security, our humanity, our faith.

At heart, I am a nineteenth-century evangelical; I was just born in the wrong century. The evangelical Christians of the nineteenth century combined revivalism with social reform and helped lead campaigns to abolish slavery and support women's suffrage and child labor laws. Charles Finney developed the idea of the "altar call" to make sure he signed up all his converts for the abolition movement. Today, poverty is the new slavery—imprisoning bodies, minds, and souls—destroying hope and ending the future for a generation.

God is acting, and the new altar call in our time is a call for faith, then a commitment—to Make Poverty History.

CHAPTER 4

The Peril

What causes wars, and what causes fightings among
you? Is it not your passions that are at war in your members? You desire and do not have; so you kill. And you
covet and cannot obtain; so you fight and wage war.

James 4:1–2

AFTER A CONTENTIOUS Methodist General Conference
meeting in the spring of 2004, an evangelical theologian,
Richard Hays, made the following proposal to his denomination, which I offer here as a revealing case study around the
issues of war and peace in the churches today.

A proposal: Let us stop fighting one another, for a season, about issues of sexuality, so that we can focus on
what God is saying to the church about our complicity
in the violence that is the deepest moral crisis of our
time. And let us call the church to fasting and prayer in
repentance for the destruction our nation has inflicted
upon the people of Iraq.

Hays continued,

The conference found ample time to debate for many
hours a stream of resolutions and actions dealing with
homosexuality, judicial procedures for charges against gay
clergy, same-sex marriage, and so forth. Not surprisingly,

the press reports of General Conference concentrated almost entirely on these "sexy" issues. . . . Am I alone in believing that we are straining at gnats and swallowing a very large camel?

Here is the moral crisis in which we find ourselves. Within the last two years, the United States launched a preemptive war, in flagrant disregard of traditional "just war" criteria, on Iraq. This military action has killed at least 10,000 Iraqis, the great majority of them civilian noncombatants. This is more than three times the number of people killed in the tragic attacks of September 11, 2001. Additionally, at this writing, about nine hundred American soldiers have died in Iraq. And these fatality counts do not begin to include the many thousands seriously wounded and maimed, on both sides of the conflict. [Updates on those figures at this writing in early September 2005 would be 1,890 American soldiers dead and at least 25,000 Iraqi civilians.]

The justifications proposed by the president and other leaders have proven false: No weapons of mass destruction have been found, Iraq was not involved in the September 11 attacks, and it had no role in sponsoring Al-Qaida. The fact that American soldiers were systematically torturing and humiliating Iraqi prisoners of war in Abu Ghraib prison is merely one painful symptom of the reckless manner in which this entire action has violated international law, despised human life, alienated allies, and fostered enmity around the world.

Despite the emergence of these facts, at no time has the president of the United States acknowledged the misguidedness of this invasion; at no time has he or any other national leader, of either political party, called the nation to apologize or repent for these violent and high-handed acts. (Instead, the religious rhetoric that we hear in the public sphere is almost always self-justifying, seldom calling us to prayerful examination of our actions and motives.) Nor—here lies the greater shame—has our church spoken out in any effective way. A few of our bishops have made clear public statements against this war, but the church as a whole has not followed their lead. Because we have remained almost completely silent, we are tacitly complicit in these actions.

As Christians, what shall we say to these things? We are called to serve a Lord who taught his followers to turn the other cheek when attacked and to love their enemies. We should also recall the passionate exhortation of St. Paul, "Beloved, never avenge yourselves, but leave room for the wrath of God. . . . Do not be overcome by evil, but overcome evil with good" (Rom. 12:19). At the heart of our call to be Jesus's disciples is our call to be peacemakers (Matt. 5:9). Yet we fail repeatedly. The gospel diagnoses our true condition: "Their feet are swift to shed blood; ruin and misery are in their paths, and the way of peace they have not known" (Rom. 3:15–17, quoting Isa. 59:7–8). And so we are called also to repentance. "If we say that we have no sin, we deceive ourselves, and the truth is not in us. If we confess our sins, he who is faithful and just will forgive our sins and cleanse us from all unrighteousness" (1 John 2:8–9). How can it be that our church has failed to call the attention of President Bush to these central teachings of the faith he professes? And how can we fail to enter into deeply penitent grief over the killing and torture that have been done in our name?

I am fully aware of the possible complexities of ethical debate over how the New Testament's teachings might apply to international affairs in a post–9/11 world. But my point is that *at present we are not having the debate at all*. In the United Methodist church, we say nothing about the horrifying violence in Iraq, while at the same time we exhaust ourselves going around in circles debating issues of sexuality.

To be sure, both sides in the sexuality debates believe that important moral issues are at stake. But, as C. S. Lewis memorably suggested in *The Screwtape Letters,* one of the devil's favorite tactics is to distract our attention with nominally good causes from the matters of first importance, so that we continue on the gradual road to hell. I am afraid that our attention has been distracted in just this way, on both sides of the sexuality debate, from serious moral reflection on the issue of war. And so we remain blind to the violence in our own hearts, a violence that surfaces in displaced form in our in-house wars over sexuality: theological conversation is supplanted by name-calling, power struggles, and manipulation of the press.

In view of the present necessity, I make the following proposal. Between now and the next General Conference in Fort Worth (2008), let us refocus our priorities. Instead of obsessively debating sexual politics, let us first devote our energies to prayerful reflection on the teaching of Jesus against violence and for peacemaking. To do this, we will have to leave some of our current debates unresolved and, nonetheless, to welcome one another as brothers and sisters. If we can join hands together and raise our voices in calling the church and the nation to prayer and repentance, it will offer a surprising and powerful witness to a world desperately needing alternatives to violence. Wouldn't it be a wonderful thing to pick up the newspaper and read, "Methodists table sex debate, call nation to repent of war"?

As precedent for this proposal, I offer Paul's counsel to the Roman Christians that they should put aside their differences over clean and unclean food—that they should stop judging and despising one another—so that they could bear witness, as Jewish and Gentile believers together, to the reconciling purpose of God. "Let us then pursue what makes for peace and mutual upbuilding. Do not, for the sake of [sex], destroy the work of God" (Rom. 14:19–20). There were sharp moral differences among the Roman Christians, serious questions about what God required. The stakes were high, but Paul refused to join the power struggle on one side or the other. Instead, he urged them, "Welcome one another, just as Christ has welcomed you, for the glory of God" (Rom. 15:7).

The affirmation of the church's unity at the conclusion of General Conference was a good starting point. But that unity will mean little unless we can bear witness together to "the gospel of peace" (Eph. 6:15). Let us then think of this next quadrennium as an extended fast, renouncing the acrimony that has clouded our discernment. The chief purposes of a fast are to clear our heads and hearts, to practice the discipline of detachment from what we suppose to be our needs, and to focus on hearing God rather than lobbying for our own agendas. Perhaps if we can take the measure of our own compromised sinfulness in matters of war and peace, we will find it possible to return to our conversation about sexuality later, and to see one another with new eyes as forgiven sinners, fellow witnesses to the one gospel of Jesus Christ in a violent and broken world.

During the past year, I have made lecturing trips to South Africa and New Zealand. In both countries, the first question repeatedly asked me by people in the churches, of all different theological stripes, was: "How can the Christians in America fail to speak out against this war?" In view of my own denomination's timid and faithless silence, I hardly knew what to answer. If we could together seek God's mercy for the chaos our nation has unleashed, I dare to hope that we might in time be able to offer a better answer to the nations, as well as to the One to whom we must ultimately render an account.[1]

I offer this extended account of the Hays proposal as a case study in misguided church priorities. Clearly, issues of sexual integrity, marriage and family, theological clarity, and church order and discipline are very important. But Hays, a conservative on the above matters, is raising a clarion cry, a prophetic warning—that we are missing what he calls the "deepest moral crisis of our time."

Hays even recognizes the complications of resolving issues of war and peace in the post-9/11 world. But he painfully points out that we are not even having a theological "debate" about such important matters. I am sure that Hays would readily admit, as would I, that Christians of good faith and conscience can and do come down on different sides of the debates over war and peace, and sometimes make conflicting prudential judgments about the facts involved to which theology must be applied.

But the central concern is that we indeed do apply theology to matters of war and peace, and not simply conform to the political, ideological, and nationalistic rationales the nations repeatedly give for going to war. In the Christian tradition, there are only three historic options for the conscientious disciple. The first option is the nonviolence of Jesus who calls each of us also to be peacemakers in a violent world. For some churches this has meant a thoroughgoing pacifism that rejects all forms of violence and even force in resolving conflicts. Others in this tradition reject the "passive" nature of that response and call for a much more active form of peacemaking and resistance to violence rather than a mere rejection of war.

The second option is the "crusade" mentality popular in the Middle Ages, when Christians went to war in God's name to defeat the "infidels" and establish Christianity, by force if necessary. Gratefully, that option has fallen out of favor in the Christian world, but is having a resurgence in modern Islamic fundamentalism and, frankly, is seeing a comeback in some of the religious rhetoric that has arisen in America's "war on terrorism."

The third option, which is the most common in American churches and the religious academies, is the "just war" with clear and rigorous criteria that must be met before war can be religiously, even if reluctantly, justified. Those criteria include that a just war can be waged only as a last resort, must be waged by a legitimate authority, must have a just cause and right intention, must have a reasonable chance of success, must be proportional to the good to be achieved, and must discriminate between combatants and civilians.

Today, in response to the twin threats of tyranny and terrorism, some are calling for perhaps a fourth way, between pacifism and just war, called "just peacemaking," which searches for "transforming initiatives" that actually do resolve conflicts, achieve justice, and make peace more possible. We will say more about that later.

There has been much discussion as to whether and how the just war tradition applies to military responses to the terrorist attacks on September 11 and, of course, to the war with Iraq. What the American media mostly failed to report was that almost every church body in the world that spoke on the Iraq war, including the Vatican and most evangelical church bodies internationally, concluded that it did not fit the just war criteria. Yet an American president with a strong personal Christian faith took the country to war anyway and American churches were left divided.

It is not the purpose of this chapter to have the Iraq debate again, and I have spoken and written on that subject in many other venues. My purpose here is simply to say, and to do so emphatically, that the issues of war and peace must be resolved *theologically* for people of faith, not merely politically. And to examine some of the biblical and theological material that must be addressed in doing so.

Most Christians would or should agree that the churches are to seek peace and work to restrain and limit the violence of the world. Whether in the pacifist or just war traditions, there is a clear Christian presumption against war. The Bible looks forward to the reign of God where both peace and justice prevail, and Christians are called to live now "as if" the kingdom of God has already broken into the world. We both look forward to the peace of God and act now in its future promise. And God's kingdom is never brought into the world by force, but rather through a cross; not through a warrior king, but a suffering servant.

Government is ordained to preserve a just and peaceful social order, restrain evil doers, and protect the common good. Although most Christians accept government's use of coercive law and force, they have historically differed on the extent of such force and their participation in it. But most would agree on the Christian preference for resolving the world's inevitable conflicts in the most peaceful ways possible. Nonviolent solutions are always to be sought and, as even just war theory rigorously requires, war must always be the last resort. In the face of modern warfare, which claims many more civilians than soldiers and easily escalates into uncontrollable conflict, these principles become even more urgent. And with the existence of weapons of mass destruction and the alarming threats of their proliferation, even to rogue states and terrorist groups, a Christian understanding of nonviolent peacemaking becomes even more urgent. Following the way of Jesus into practical peacemaking and nonviolent conflict resolution has become a critical vocation for Christians in a world of violence.

THE WORLD OF VIOLENCE causes a hardness of heart, which affects all aspects of our political life. It is the moral price we have paid for the normalcy of modern warfare and the acceptance of weapons of mass destruction. Biblically speaking, hardness of heart is not so much the deliberate doing of evil. Rather, it is the loss of the ability to distinguish between good and evil. It is like having deadened nerves that cannot feel. It means having eyes that no longer see and ears that no longer hear. Reason becomes clouded and hearts closed. Hardness of heart results in a callousness to the pain of others. The official political debates and military policy discussions in Washington and around the world reveal that hardness of heart. We close our hearts to understanding and even seeing the suffering of the people in circumstances of conflict and war. We are cut off from the human reality of what the next use of our weapons and war-fighting strategies will be like. With hardened hearts, we feel neither the victims' pain nor our own conscience; all we feel is fear, confusion, and resignation to "the realities of war."

None of us is immune to this hardness of heart. All of us are deeply affected by it. No longer can any of us self-righteously say, "They are wrong, and we are right." To turn our hearts, minds, and bodies against war and toward peace can be done only in response to faith. When we recognize our own spiritual condition, we realize that we don't want our hearts to remain hard any longer. Then we act so that we might have our hearts softened. Our work for peace is done so that we might be reminded of God and of our neighbor whom we have resigned to destruction. We need to be sensitized, opened up, and changed so we can raise our voices, such as they are, in the hope that others might also have their hearts softened.

The antidote in Scripture for hardness of heart is conversion. The only thing that finally can soften hearts is the presence of Christ among us, which reveals how numb we have become. The children of God are to have their hearts softened by their Lord so that they might be peacemakers. Their hearts have been opened so that they might begin to move in a different direction.

The Bible speaks with compelling urgency to the situation we now face. The words of the book of Deuteronomy are for us:

See, I have set before you this day life and good, death and evil. If you obey the commandments of the Lord your God which I command you this day, by loving the

Lord your God, by walking in his ways, and by keeping his commandments and his statutes and his ordinances, then you shall live and multiply, and the Lord your God will bless you in the land which you are entering to take possession of it. But if your heart turns away, and you will not hear, but are drawn away to worship other gods and serve them, I declare to you this day, that you shall perish; you shall not live long in the land which you are going over the Jordan to enter and possess. I call heaven and earth to witness against you this day, that I have set before you life and death, blessing and curse; therefore choose life, that you and your descendants may live. (Deut. 30:15–19)

The Old Covenant set limits on the human propensity for violence; its law was meant to restrain the scope of retaliation and bloodshed. The prophets repeatedly attacked the roots of violence in greed, envy, hate, and self-righteousness. The Old Testament thus outlaws the militarism practiced by our own nation and others today. Hosea points out the consequences of trusting in chariots and warriors.

Because you have trusted in your chariots and in the multitude of your warriors, therefore the tumult of war shall arise among your people, and all your fortresses shall be destroyed. (Hos. 10:13b–14a)

The Israelites were not to put their hopes in military weapons and alliances but were to rely upon Yahweh for their protection.

Woe to those who go down to Egypt for help and rely on horses, who trust in chariots because they are many

and in horsemen because they are very strong, but do
not look to the Holy One of Israel or consult the Lord!
(Isa. 31:1)

The prophets Isaiah and Micah make clear that God's will
is not for war but for peace, and that someday this peace will
be established among the nations.

It shall come to pass in the latter days that the mountain
of the house of the Lord shall be established as the highest of the mountains, and shall be raised up above the
hills; and peoples shall flow to it. . . . For out of Zion
shall go forth the law, and the word of the Lord from
Jerusalem. He shall judge between many peoples, and
shall decide for strong nations afar off; and they shall
beat their swords into plowshares, and their spears into
pruning hooks; nation shall not lift up sword against nation, neither shall they learn war any more. (Mic. 4:1–3)

Throughout the Old Testament, prophecies point to a
New Covenant and to One who will guide the people in the
ways of peace. In the well-known words of Isaiah foretelling
the birth of Christ, the prophet culminates his litany of appel-
lations with the title "Prince of Peace." The peace he will establish, says Isaiah, will never end (Isa. 9:6–7).

The ministry of John the Baptist was to prepare the way of
the Lord, "to give knowledge of salvation to his people . . . to
give light to those who sit in darkness . . . to guide our feet
into the way of peace" (Luke 1:77–79). An angel of the Lord
appeared to shepherds in the field to announce the birth of
the Savior. "And suddenly there was with the angel a multitude of the heavenly host praising God and saying, 'Glory to

God in the highest and on earth peace, goodwill among men'" (Luke 2:13–14). The coming of Jesus meant the coming of peace.

Jesus would show another way to live. No longer would violence be limited to returning no more than was done to you; violence would be met with an entirely different response. The principle of equal retribution would be replaced with the practice of suffering love.

> You have heard that it was said, "An eye for an eye and a tooth for a tooth." But I say to you, Do not resist one who is evil. But if any one strikes you on the right cheek, turn to him the other also; and if any one would sue you and take your coat, let him have your cloak as well; and if any one forces you to go one mile, go with him two miles. Give to him who begs from you, and do not refuse him who would borrow from you. You have heard it said, "You shall love your neighbor and hate your enemy." But I say to you, Love your enemies and pray for those who persecute you, so that you may be sons of your Father who is in heaven. (Matt. 5:38–45)

With these words, Jesus placed an obstacle in the way of war. This obstacle has been repeatedly pushed aside by the nations as they find endless justifications for hating and making war with one another.

"Blessed are the peacemakers," said Jesus, "for they shall be called children of God" (Matt. 5:9). Many will succumb to fear, accede to prejudice, join the hysteria, and rush to war; but others will remember the way of their Lord and seek other means for resolving human conflicts. Those who speak up for peace are God's very own. They remember that they too were once enemies of God, and they recall how God dealt with them: not by destroying them, but by reconciling them to himself and to one another. This is God's way, and to be God's children means to follow in God's way.

God's intention for reconciliation through Jesus Christ is made clear in the New Testament. No longer merely restrained in our violence, we are now called to "be merciful, even as your Father is merciful" (Luke 6:36), and to imitate Christ in his way of peace. Jesus's earliest followers were ready to die, but not to kill. And they learned that making peace is always costly. Jesus made peace, and disarmed the principalities and powers, at the cost of his life (Col. 1:19–20; 2:13–15).

Peace is won at the price of Jesus's own blood, and the cross becomes the sign of God's reconciliation. God has overcome evil, not by means of a sword but through the cross. Violence has been defeated and death has lost its sting, not through greater violence but by the power of love and suffering. All this has cosmic meaning—the great forces of the world are all involved—as well as personal meaning. The reign of sin has been broken, the powers of the world have been vanquished, and all things, "whether on earth or in heaven," have been reconciled because of One who was willing to make peace by giving his life. The great work of reconciliation, in all its personal and political dimensions, was accomplished in the cross of Christ.

Those who had been at odds have been reconciled to one another. In and through Christ, God has created a new people. It includes our friends (those similar to us) and it includes our enemies (those different and divided from us by race, class, sex, or national identity). Old hostilities have come to an end for those in Christ Jesus.

But now in Christ Jesus you who once were far off have been brought near in the blood of Christ. For he is our peace, who has made us both one, and has broken down the dividing wall of hostility, by abolishing in his flesh the law of commandments and ordinances, that he might create in himself one new man in place of the two, so making peace, and might reconcile us both to God in one body through the cross, thereby bringing the hostility to an end. (Eph. 2:13–16)

If we belong to Christ, the work of reconciliation is our vocation. Jesus said, "If any man would come after me, let him deny himself and take up his cross and follow me" (Mark 8:34). Jesus continually told his disciples that sharing his life means sharing his suffering. His way inevitably leads to Jerusalem and to conflict with the world. All along the way the disciples resisted. They sought to avoid the implications of Jesus's vocation and, ultimately, of their own. Paul confirms in his epistles that, indeed, to follow Jesus means to follow the way of the cross. We are called to imitate Jesus, not in his carpentry nor in his celibacy but in his sufferings. There will be no glory without suffering, says Paul. To share in Jesus's glory the disciples must be prepared to share in his way, and his way is the cross. "He who does not take up his cross and follow me is not worthy of me" (Matt. 10:38). Dietrich Bonhoeffer understood the point when he said, "When Jesus calls a man, he bids him come and die."[2] We have no reason or right to choose another way than the way God chose in Jesus Christ. The cross is both the symbol of our salvation and the pattern for our lives. We are joined to Christ in his ministry of reconciliation, which has now been entrusted to us.

> Therefore, if any one is in Christ, he is a new creation;
> the old has passed away, behold, the new has come. All
> this is from God, who through Christ reconciled us to
> himself and gave us the ministry of reconciliation; that
> is, God was in Christ reconciling the world to himself,
> not counting their trespasses against them, and entrust-
> ing to us the message of reconciliation. (2 Cor. 5:17–19)

The violence of the nations increases daily. More and more
of the resources of the world are being used to make preparation for war. The forces of violence and the powers that be
have placed the world in unprecedented danger. War is ever
impending, and weapons of mass destruction are abundant.
And the people of God are quiet and acquiescent. The churches are
immobilized, unable to name the idolatry and see the heresy.
Protecting ourselves from fear, fighting for our own ideology,
defending our national self-interest, protecting our standard of
living: these have become the justifications for making war;
these have become our gods. In the churches, as in the political arena, most of the current debate over war is constricted
to paying homage to these gods. If the churches continue to
agree with the politicians about the acceptable boundaries of
the debate, the cause of peace will forever be diminished.

A church that places its trust in weapons and war is a
church that no longer trusts in the Lord. We live in nations
where we are told that to be secure we must place our faith in
systems capable of destroying millions and literally incinerat-
ing the globe. Can we possibly agree with that and still say,
"We are God's people, we belong to the Lord, and we place
our security only in him"? The God of the Bible hungers for
peace and reconciliation among all men and women. Are we
willing to move into the midst of the world's violence as

agents of that reconciliation? The question at stake in a world awash with weapons of mass destruction is whether the church will demonstrate to whom it ultimately belongs. A church that is accommodated to the logic of war is making a statement about its own false worship.

Some religious leaders suggest that our nation is better and more blessed than others; they proclaim that our military purposes are righteous and just. Our enemies alone are perpetrators of evil; we are defenders of freedom. Our pursuit of military supremacy becomes, by this very American logic, the key to peace in the world. The prospect of endless war, even with weapons of mass destruction, generates little reflection; rather, it takes on the character of a moral crusade. Some of our preachers on the Religious Right even imply that the church needs the protection of the American military umbrella in order to wage war against evil in the world.

The new enemy of terrorism has proven quite resilient against mere military responses. More effective would be to undermine the ability of the terrorists to recruit for their murderous agendas by resolving the conflicts, dealing with the grievances, and changing the policies that fuel the angers and resentments that sustain the terrorists and their evil mission.

If we say, and we must as clearly as we can, that there is never any justification or even explanation for the hideous kind of terrorist attacks on civilian populations we now see around the world, then we must never ourselves engage in the double standard of attacking more innocents in our pursuit of the terrorists. The world is full of grievances, real and imagined. But the taking of innocent life in response to those grievances is never morally acceptable. Never. That is the message the terrorists need to hear, but in our response to terrorist attacks, we must also not commit the moral offense of taking innocent lives. The strategy of unilateral and preemptive military strategies to counter terrorism and, indeed, to go on the "offensive" has become both a moral and political mistake that leads to many unintended consequences that can actually make the situation worse.

If war is failing to solve our deepest problems, resolve our hardest conflicts, and make us safe, what is the alternative? Peacemaking must be practical as well as faithful. To speak of successful peacemaking and conflict resolution is not to engage in utopian thinking but to find what actually works to limit and ultimately replace violence. Many people are being drawn to such practical and successful nonviolent strategies. Here is a summary of the best contemporary thinking on "just peacemaking" offered by a remarkable group of ethicists (from both the pacifist and just war traditions) led by Fuller Seminary professor of ethics, Glen Stassen. They are called "Ten Practices for Abolishing War."[3]

1. *Support nonviolent direct action.* Nonviolent direct action is spreading rapidly, ending dictatorship in the Philippines, ending rule by the Shah in Iran, bringing about nonviolent revolutions in Poland, East Germany, and Central Europe, transforming injustice into democratic change in human rights movements in Guatemala, Argentina, and elsewhere in Latin America, in South Africa. . . . Governments and people have the obligation to make room for and to support nonviolent direct action.

2. *Take independent initiatives to reduce threat.* Independent initiatives: (1) are independent of the slow process of negotiation; (2) decrease threat perception and distrust but do not leave the initiator weak; (3) are verifiable actions; (4)

and carried out at the announced time regardless of the other side's bluster; (5) have their purpose clearly announced—to shift toward de-escalation and to invite reciprocation; (6) come in a series; initiatives should continue in order to keep inviting reciprocation. This new practice has been crucial in several recent breakthroughs.

3. *Use cooperative conflict resolution.* (1) Active partnership in developing solutions, not merely passive cooperation. (2) Adversaries listen to each other and experience each others' perspectives, including culture, spirituality, story, history, and emotion. (3) Seek long-term solutions that help prevent future conflict. (4) Seek justice as a core component for sustainable peace.

4. *Acknowledge responsibility for conflict and injustice and seek repentance and forgiveness.* Until recently, it was widely agreed that nations would not express regret, acknowledge responsibility, or give forgiveness. But Germany since World War II, Japan and Korea, Clinton in Africa, the U.S. finally toward Japanese-Americans during World War II, the South African Truth and Reconciliation Commission, and other actions described by Shriver, *An Ethic for Enemies,* and Wink, *When Powers Fall,* show a crucial new practice is emerging that can heal longstanding bitternesses.

5. *Advance democracy, human rights, and religious liberty.* Extensive empirical evidence shows that the spreading of democracy and respect for human rights, including religious liberty, is widening the zones of peace. Democracies fought *no wars* against one another during the entire twentieth century. They had fewer civil wars. And they

generally devoted lower shares of their national products to military expenditures, which decreases threats to other countries.

Ties of economic interdependence by trade and investment also decrease the incidence of war. Engagement in international organizations like the UN and regional institutions is a clear predictive factor that they will be much less likely to engage in war.

6. *Foster just and sustainable economic development.* Sustainable development occurs where the needs of today are met without threatening the needs of tomorrow—where those who lack adequate material and economic resources gain access, and those who have learn to control resource use and prevent future exhaustion.

A key to economic development in East Asian countries, especially [South] Korea and Taiwan, has been land reform that made wealth more equitable and thus created a sizable local market for developing firms. By contrast, Latin America lacks real land reform and equality, and therefore local consumers cannot afford to buy products produced by local industries.

7. *Work with emerging cooperative forces in the international system.* Four trends have so altered the conditions and practices of international relations as to make it possible now, where it was not possible before, to form and sustain voluntary associations for peace and other valuable common purposes that are in fact working: the decline in the utility of war; the priority of trade and the economy over war; the strength of international exchanges, communications, transactions, and networks; and the gradual ascendancy of liberal representative democracy and a

mixture of welfare–state and laissez–faire market economy. We should act so as to strengthen these trends and the international associations that they make possible.

8. *Strengthen the United Nations and international efforts for cooperation and human rights.* Acting alone, states cannot solve problems of trade, debt, interest rates; of pollution, ozone depletion, acid rain, depletion of fish stocks, global warming; of migrations and refugees seeking asylum; of military security when weapons rapidly penetrate borders.

   Therefore, collective action is increasingly necessary. U.S. citizens should press their government to pay its UN dues and to act in ways that strengthen the effectiveness of the United Nations, of regional organizations, and of multilateral peacemaking, peacekeeping, and peace building. They resolve conflicts, monitor, nurture, and even enforce truces. They meet human needs for food, hygiene, medicine, education, and economic interaction. Most wars now happen within states, not between states; therefore, collective action needs to include UN–approved humanitarian intervention in cases like the former Yugoslavia, Haiti, Somalia, and Rwanda "when a state's condition or behavior results in ... grave and massive violations of human rights."

9. *Reduce offensive weapons and weapons trade.* A key factor in the decrease of war between nations is that weapons have become so destructive that war is not worth the price. Reducing offensive weapons and shifting toward defensive force structures strengthens that equation. Banning chemical and biological weapons, and reducing strategic (long-range) nuclear warheads from 3,500 to 1,000 each, are key steps.

Arms imports by developing nations in 1995 dropped to one-quarter of their peak in 1988. But the power of money invested by arms manufacturers in politicians' campaigns is a major obstacle to reductions.

10. *Encourage grassroots peacemaking groups and voluntary associations.* The existence of a growing worldwide people's movement constitutes one more historical force that makes just peacemaking theory possible. They learn peacemaking practices and press governments to employ these practices; governments should protect such associations in law, and give them accurate information.

Each practice is recent in its widespread use, and is causing significant change. Together they exert strong influence, decreasing wars. Each is empirically happening and being effective in abolishing some wars. Each faces significant obstacles and blocking forces that are named in the chapters. We contend that just peacemaking practices are ethically obligatory for persons, groups, and governments to strengthen them and help overcome the blocking forces. But at the deepest level, peacemaking will not just depend upon good strategies but on our own real conversion.

PRAYER. THAT IS THE WORD that kept coming back to me while attending a conference of activists many years ago. The word seemed a little incongruous at first. There I was, in the midst of a meeting to plan scenarios for political action, and my strongest sense was the need for prayer. But that word was soon confirmed. Other Christians at the conference approached me individually and expressed their need to pray.

These initial feelings soon deepened and developed into a specific sense of being called to prayer as an active response to the danger of war. That call was strongly confirmed in the Sojourners community, and created among us a heightened expectation of God's leading, and a clearer sense of our own direction in peacemaking.

When the forces of destruction at work in the world appear so overwhelming, and when we often feel so hopeless and helpless, there is nothing more important for Christians than to be moved to and moved by prayer. The beginning of our response to great evil is to go to our knees. The threats of terrorism, war, and weapons of mass destruction now hanging over the world should drive Christians to a new depth and intensity of prayer. In the Bible, prayer is not the passive, general, and abstract sort of ritual common in many of our churches. Biblical prayer is more active, concrete, historically specific, and politically informed. Prayer in the Bible is offered *for* some things and *against* other things. Prayer is for salvation, healing, peace, righteousness, justice, deliverance, and protection. Prayer is against sin, unrighteousness, injustice, idolatry, destruction, and calamity. Prayer is directed against kings, rulers, and leaders when their behavior is evil and contrary to the welfare of the people.

In the sixth chapter of Ephesians, Paul says we are engaged in spiritual warfare with the principalities and powers, and he lists prayer as a weapon to be used against those powers. Today prayer must be used against the powers of violence that have us in their grip. When Paul speaks of Christ's disarming of the powers, he simply means that Jesus exposed their lie, showed them for what they were, unmasked the illusion of their power, and stood free of their rewards and punishments. Jesus's freedom from the fear and control of the powers was rooted in the deep knowledge of who he was and to whom he belonged. His communion with God was his constant source of strength and power.

Prayer is the act of reclaiming our identity as the children of God; it declares who we are and to whom we belong. The action of prayer places us outside the realm of the powers and principalities. As prayer declares our true identity, it destroys our false identities. In prayer we act upon who we really are, and thus prayer has the effect of diminishing the illusions that have controlled us. It is therefore an act of revealing the truth and unmasking the lie. Prayer allows us to step out of our traps and find ourselves again in God.

Spiritual writer Henri Nouwen once shared with our community how the desert fathers regarded prayer as an act of "unhooking" from the harness of the world's securities. Such prayer may be the only action powerful enough to free us from our spiritual bondage to the fear, property, money, power, ideas, and causes that have made us willing to destroy so much and so many in the desperate and futile effort to protect ourselves. Only those who have truly found their security in God can resist the violent tugs and pulls of the false securities offered by the powers. By reestablishing our security in God, prayer becomes an effective weapon in combating those powers.

Historically, prayer and worship have always been at the heart of the most powerful expressions of Christian witness against tyranny and violence. Prayer, while offered for the sake of the world, will change those who pray. Motivated by a great evil in the world, prayer first raises the question of our own complicity in the evil. Prayer humbles us. It starts in confession and repentance and recalls our identity as God's people. To pray is to recognize that, before the evil can be overcome, we must be transformed. Prayer, therefore, is central to conversion. Prayer, in recognizing God's authority over the powers, moves us beyond opposition to affirmation, beyond resistance to celebration. Thus prayer and the results of prayer are the most revolutionary of acts. The powers and principalities of this world are aware of this; that is why they consider those who pray in this way to be a threat.

Prayer changes our frame of reference; it is not merely a preparation for action. Prayer must be understood as an action in itself, a potent political weapon to be used in spiritual warfare against the most powerful forces of the world. Prayer is not undertaken in place of other actions; it is the foundation for all the other actions we take.

Given the enormity of the present dangers of violence and war, prayer is a necessity. Without it we see only our point of view, our own righteousness, and ignore the perspective of our enemies. Prayer breaks down those distinctions. To do violence to others, you must make them into enemies. Prayer, on the other hand, makes enemies into human beings. When we have brought our enemies into our hearts in prayer, it becomes most difficult to maintain the hostility necessary for violence. In bringing them closer to us, prayer serves to protect our enemies. Thus, prayer undermines the propaganda and policies of governments designed to make us hate and fear our enemies. By softening our hearts toward our adversaries, prayer can become treasonous. Fervent prayer for our enemies is a great obstacle to war and the feelings that lead to it.

Consider what might happen if the churches made prayer for our enemies a regular part of the Eucharist. If, when we prayed for unity, we included a prayer for our enemies, if every time we gathered for worship, we paused to remember and pray for the particular people our governments have termed "enemies"—how differently might we begin to re-gard our adversaries? Our whole posture toward them would change. In a world full of such destructive weapons, no change could be more welcome or urgent. Particularly in an age of real terrorist threats and the temptation to respond in revenge, prayers for our enemies should be incorporated into the Eucharistic liturgy. Such a practice would have the effect of confronting every Christian with the humanity of those who oppose us and even attack us, and it would do so right at the heart of our worship of Christ. Such a proposal should be offered in every church and denomination.

Prayer for our enemies takes them into our hearts. It would bring potential adversaries into our daily consciousness, serv-ing as a regular reminder of their existence. As we pray, we begin to see people as God sees them. With continuing prayer, our fear subsides, our anger and hurt is gradually healed, and we begin to understand the others' fears, angers, and hurts. It is exceedingly difficult to hate people while we are praying for them. Prayer undermines hostility and enables us to identify with another person. We bring them to mind in the presence of God, and our minds are changed toward them. We begin to understand things from their perspective: their feelings, fears, pains, struggles, joys, and hopes. To iden-tify with an enemy is to turn an enemy into a neighbor.

At present, however, our enemies remain faceless to us. Facelessness is, in fact, their prime characteristic. We assure ourselves that they are nothing like us. We are good and they are bad. Our cause is noble, but their purpose is evil. We can be trusted but they are completely untrustworthy. They know neither reason nor love, only irrationality and hate. It is almost impossible to imagine our enemies showing kindness, gentle-ness, or compassion. Have you ever wondered why we almost never picture our adversaries with their children? To do so would be to see the intimacy we know in ourselves but cannot accept in our enemies.

Necessary to the preparation for war is the dehumanization of our enemies. They become sinister and disgusting. Our enemy's way of life is depicted as inhuman. They are seen bearing nothing but ill intentions toward us. We tag them with derogatory names that further hide the human faces of those we are making ready to destroy. Japanese become "Japs" or "Nips," Germans become "Krauts" or "Jerrys," Vietnamese become "Gooks," "Slopes," or "Slant Eyes," and, of course, Russians were simply "Communists" and Muslims are now "terrorists." Lt. William Calley, on trial for the murder of Vietnamese villagers at My Lai, said in his own defense, "In all my years in the Army, I was never taught that communists are human beings."

Since our enemies are incapable of normal human feelings and values, they are feared. Their very existence threatens our way of life. Since they are less than human, they must be treated as such. They are different, so they must be dealt with differently. Force is the only thing that people like these understand. It is the only choice we have—War.

"The root of war is fear," said Trappist monk Thomas Merton. Conversion, including the renewal of compassion for our neighbor, is the only enduring road to peace. Understanding and defending the humanity of enemy populations becomes increasingly important as pressures for war mount. Reviving our capacity to love has become an urgent political necessity as the nations come to regard millions of their neighbors as nothing more than expendable enemy populations in war. We face unimaginable destruction unless our hearts are enlarged to recognize a neighbor in the face of our enemy.

This kind of conversion is not likely to happen overnight. We have already discussed how it is most difficult for affluent Americans to see the face of poverty. But if it is hard for us to establish proximity to the poor who are all around us, it is even harder to touch the human faces of the victims of war. Just as conversion in a hungry world means coming face-to-face with the poor, conversion in an age of violence and counterviolence means coming face-to-face with the human reality of the many victims. In each case, the converted person is one who can see and feel the human suffering. In other words, to be converted is to have compassion for the victims. It means to carry, in your own heart, the human suffering of others. During the Vietnam War, the peace efforts that were most enduring usually came from those persons who deeply felt the agony of Vietnamese people. Those whose actions were more ideologically motivated failed to produce the same quality of peace activity. The same is true today in people's response to the pain of Rwanda, the Balkans, the Middle East, or Darfur, Sudan. In every case, it is the human face of suffering that begins to melt our hardness of heart.

When war comes, we all see and are horrified. But then it is too late, and we wonder, in our agony, why we did not see earlier. Conversion is to see the human face of war *now*, before it occurs. Conversion means to enter into anticipatory suffering with every potential victim. To see is to act. We cannot hide our faces from the poor; neither can we turn away from the faces of the victims of violence. In both cases, to turn away from them is to turn away from our own flesh. In a hungry world, we suffer the loss of what children might have otherwise become. In a world of violence and war, we suffer the loss of any future for children at all. Our blindness must be removed and our hearts opened to the reality. This too is a radical surgery for which we must also turn to the Lord. Conversion always means two things: seeing Jesus and seeing our neighbor. Perhaps when we see Jesus in the faces of the victims of war, our hearts will be opened.

On the cross, Jesus took upon himself every sin, every hate, every fear, every violence, and every death, including those of every war. He took our place. He represented us. Jesus bore in his own body the wrath that we deserved. The crucified God bore the pain of all the victims of war. He was there with them. By remembering the weight of the sin he bore at Calvary, we have an intimation of his agony at the world's continuing wars. By seeing him there, we begin to understand the sufferings of those places.

Perhaps we can better comprehend the deaths in war if we realize that Jesus is the central victim.* He knows each victim. Each hair on every head he has numbered. Each passion, fear, and love is familiar to him. Jesus is there with every father, mother, and terrified child. He feels every death.

The word that must go forth from the church is the first word from Jesus's own sermon: *Repent.* However, to stop, turn around, and go in a different direction will be very difficult without an alternative, without visible demonstration of how we might live in our perilous age. To change how we live will require more than good preaching and fine statements. There have been endless appeals from the churches and from ecumenical bodies on war. The problem with the appeals is that they have usually centered upon what Christians think other people should do rather than on what Christians will do. The

---

* I am indebted to Dale Aukerman for this insight of Jesus as the central victim of war. See his book, *The Darkening Valley: A Biblical Perspective on Nuclear War* (New York: Seabury Press, 1981).

emphasis has been on what the government ought to do or what other citizens should do. Very few of the Christian statements have said, "This is what we intend to do, and, this is what we refuse to do."

Anyone can love peace, but Jesus doesn't say, "Blessed are the peacelovers." He says peace*makers*. He is referring to a life vocation, not a hobby on the sidelines of life. And he means a vocation that includes risk. Too often, the advocates of pacifism today are more noted for the things they won't do than for the things they are willing to suffer for the cause of peace. And the advocates of "just war" often find too many wars acceptable. The making of peace, as Jesus showed in his own life, can result in great conflict. Such a ministry of reconciliation will cost something, and it will often make us misunderstood in a world that knows violence better than peace.

Converts must become known as peacemakers. We must follow the leadership of the One who was willing to bear the cost of making peace in a hostile world. To risk peace is to state that the risks involved in our present course of endless war are much greater. In the midst of preparations for war that never ends, the church must dare to demonstrate another way. No longer trusting in our weapons for security, we must learn to trust in the things that make for peace. And trusting anew in God, we must begin cooperating with one another in preparations for peace.

Jesus put himself in the breach, in the midst of the conflict and the violence. In every situation of violence, the church should ask: What does it mean, in this situation, to put ourselves in the midst of the conflict, to stand between the contending forces? Its answer will tell us how we take on the vocation of peacemaking. That vocation always entails risk, whether the risk is our own or that which, through our support, we enable a sister or brother to take. It means experimenting with the victory of Christ in our lives.

We also need to remember that communication is synonymous with nonviolence. Our best actions are those which create situations that make dialogue more possible. Even when dramatic action is necessary, the action must still have the purpose of creating dialogue. Our forms of action must be continually examined and corrected so that we don't alienate or confuse the very people we seek to reach. Rather, we need to act in a way that includes, invites, clarifies, and gains the confidence and respect of those who see and hear what we have to say.

While recognizing the enormous dangers of violence and the painful risks of conversion, we must refuse to accept despair or to act out of it. Conversion is always possible as long as there is life. Our prayer must be for patience, and for the insight to demonstrate concrete, positive alternatives that show people what we are *for*. The quality of what we communicate is most important. Is our message marked by hope, by truth, by courage? Or does it convey despair, contempt, bitterness, and frustration? Do we despair of people changing? If we do, then they probably won't.

It is important to note that where the reality of apocalypse is most pronounced in the Bible, it is in the same place and in the same people that the power of worship and praise is most evident. Those who are most truly conscious of the apocalyptic character of their times will also be those who have learned what it means to worship their God and to celebrate the Word of God in their lives and in their historical circumstances.

We must never lose sight of the victory of Christ in the face of violence. That victory in history is assured. It has already been won for us. We must learn to be a people who can see the world in the light of the victory of Christ. Prayer can help us to know that victory as real, to evaluate all things by it, and to act upon it. If we know the victory of Christ to be true in our experience, we can demonstrate it and manifest it beyond the boundaries of our communities for the world to see.

His victory is the basis for our hope. I am convinced that the appeal to war will not be overcome by an appeal to fear. Its own basis is fear, and more fear will not finally prevail against it. Rather, violence will be overcome with hope. Christian hope sees the world situation realistically, with no false optimism. But Christian hope knows that the victory of Christ is stronger than the powers of violence. Prayer helps us to continue to believe that. Our actions must show the world that we believe.

The life of Christ, which has been planted in us and among us, is stronger than the forces around us. It is not that "we shall overcome"; it is rather that this is the day of the Lord, and he shall overcome. Despite the pretentions of the powers of violence, the Lamb who was slain has begun his reign. Hallelujah!

CHAPTER 5

The Vision

Through the church the manifold wisdom of God [is] now made known to the principalities and powers in the heavenly places. This was according to the eternal purposes which [God] has realized in Christ Jesus our Lord....

Ephesians 3:10–11

WHEN I WAS A university student, I was unsuccessfully evangelized by almost every Christian group on campus. My basic response to their preaching was, "How can I believe when I look at the way the church lives?" They answered, "Don't look at the church—look at Jesus."

I now believe that statement is one of the saddest in the history of the church. It puts Jesus on a pedestal apart from the people who name his name. Belief in him becomes an abstraction removed from any demonstration of its meaning in the world. Such thinking is a denial of what is most basic to the gospel: incarnation. People should be able to look at the way we live and begin to understand what the gospel is about. Our life must tell them who Jesus is and what he cares about.

Everyone today seems to recognize that we have serious problems in the church. There is wide agreement that fundamental change needs to take place. But there the agreement ends. A variety of answers is offered. For many, the solution to our problems is the preaching of the Word—renewal through evangelism. For others, it is the filling by the Holy Spirit—renewal through charismatic gifts. For some, it is service to the poor and political action on behalf of justice. And for others, it is acts of resistance to the power and violence of the state. All of the answers are right. Each speaks to a glaring lack in the church's life, and each contributes to a fuller understanding of what the church is meant to be in the world. But all of the answers are inadequate.

The greatest need in our time is not simply for *kerygma,* the preaching of the gospel; nor for *diakonia,* service on behalf of justice; nor for *charisma,* the experience of the Spirit's gifts; nor even for *propheteia,* the challenging of the king. The greatest need of our time is for *koinonia,* the call simply to be the church, to love one another, and to offer our lives for the sake of the world. The creation of living, breathing, loving communities of faith at the local church level is the foundation of all the other answers. The community of faith incarnates a whole new order, offers a visible and concrete alternative, and issues a basic challenge to the world as it is. The church must be called to be the church, to rebuild the kind of community that gives substance to the claims of faith.

To be the church is first to know the biblical identity and vocation of the community of faith. Throughout this book we have stressed the importance of the church as a community created by conversion and offered for the sake of the world. Here we will develop a more detailed biblical picture of what this means.

Behold my servant, whom I uphold, my chosen, in whom my soul delights; I have put my spirit upon him, he will bring forth justice to the nations. He will not cry or lift up his voice, or make it heard in the street; a bruised reed he will not break, and a dimly burning wick he will not quench; he will faithfully bring forth justice.... "I am the Lord, I have called you in righteousness, I have taken you by the hand and kept you; I have given you as a covenant to the people, a light to the nations, to open the eyes that are blind, to bring out the prisoners from the dungeon, from the prison those who sit in darkness. I am the Lord, that is my name; my glory I give to no other, nor my praise to graven images. Behold, the former things have come to pass, and new things I now declare; before they spring forth I tell you of them." (Isa. 42:1–3, 6–9)

The "servant songs" of Isaiah foretell the coming of the Lord's anointed one. They are a prophetic picture of how God's purposes will be carried out in history. The vehicle for God's saving purposes will be the one known as "the suffering servant." This is the one whom God has chosen and in whom the Lord delights. The passage both affirms that relationship and sets forth the mission of God's anointed. Isaiah links the anointing of the Spirit with a threefold repetition of God's call to establish justice in the world. The Lord's servant has been endowed with the Spirit for that very purpose, to "bring forth justice to the nations." In other words, the coming of the Spirit is directly connected with the accomplishing of God's purposes of justice, not just within the believing community, but throughout the world.

The servant's speech will be quiet and humble. His manner will be gentle and without violence. He will be careful not to break what is already weak or quench what is only flickering. But his faith will be focused and his direction clear. He will never waver nor be dissuaded. The servant of the Lord will be unmistakable in his purpose, unshakable in his resolve, and unrelenting in his persistence "til he has established justice in the earth." Isaiah says the world is anxiously awaiting his arrival. Already, in the prophecy, there is an inkling of a corporate as well as an individual identity for the suffering servant. The second part of the passage suggests the vocation of a people and not just a person.

"I am the Lord, I have called you in righteousness, I have taken you by the hand and kept you." The Lord's servant has not been called to a righteous purpose and sent off alone; he has been taken by the hand and cared for by the one whose purpose he serves. There is an intimacy to these words, a warm invitation to relationship. The relationship between the servant and his Lord is the beginning and the foundation of establishing justice in the nations. Isaiah goes on to tell how the suffering servant will lay down his life in order that God's will be accomplished.

At his baptism, the Spirit of God identified Jesus as the beloved Son of God.

> Then Jesus came from Galilee to the Jordan to John, to be baptized by him. John would have prevented him, saying, "I need to be baptized by you, and do you come to me?" But Jesus answered him, "Let it be so now; for thus it is fitting for us to fulfil all righteousness." Then he consented. And when Jesus was baptized, he went up immediately from the water, and behold, the heavens were opened and he saw the Spirit of God descending like a dove, and alighting on him; and lo, a voice from heaven, saying, "This is my beloved Son, with whom I am well pleased." (Matt. 3:13–17)

Jesus's baptism designated him as the one of whom Isaiah spoke. In and through him, the purposes of God would be fulfilled. Jesus's self-consciousness as God's suffering servant is revealed in his Nazareth manifesto (Luke 4:17–21), to which we have already referred. He had gone to the temple, opened the book, and read from Isaiah 61, another of the servant songs. He then closed the book, gave it back to the attendant, and sat down. All eyes were fixed on him, and he said to them, "Today this scripture has been fulfilled in your hearing" (Luke 4:21). His anointing by the Spirit was for the purpose prophesied by Isaiah—to establish justice in the earth—and Jesus spelled that out by identifying the poor, the captives, the blind, and the oppressed as central to his mission. Proclaiming "the acceptable year of the Lord" likely refers, as we have said, to the Jubilee tradition of economic redistribution. Jesus's words in his Nazareth sermon were to be the constant themes of his ministry. The key, in Isaiah 42 and 61, Matthew 3, and Luke 4, is the integral connection between the anointing of the Spirit, the identity of the suffering servant, and God's purposes of justice in the world.

The baptism of the church at Pentecost is for these same purposes. The mission prophesied by Isaiah and embodied in Jesus is fulfilled in the company of his followers. They are "the body of Christ," and together bear the same anointing as he. Their corporate life is a continuation of the vocation of Jesus in the world. The Christian fellowship is, therefore, "the suffering servant community." The beginning of its anointing is recorded in Acts 2, the consequences of which we have already discussed.

From the outset, the life of the Christian community is dependent upon the coming of the Holy Spirit. Here again, the connection between the anointing and the carrying out of

God's purposes is crucial. Just as the baptism of Jesus identified him as the one with whom God was pleased and whom God would use, so the baptism of the Spirit at Pentecost identifies the church as the community beloved of God and created by the Holy Spirit for the purposes of Christ in the world. Such a life means that believers are bound together as never before. They are brothers and sisters in the family of God and have become united in a deep love and a common task. The vocation of God's suffering servant had been embodied perfectly in one life, but now that same vocation can be seen, with the same authority, in the body of Christ as its members experience the gifts of God's Spirit.

The vocation and the very identity of the Christian community, therefore, is directly in line with Isaiah's description of the suffering servant. The same servant posture and style we saw in Isaiah and witnessed in Jesus characterizes the believing community. We are a people drawn into relationship to the Lord for the sake of God's purposes in history. The body of Christ has been anointed and empowered for God's mission to bring justice and reconciliation. The followers of Jesus, after his example, will lay down their lives for the kingdom; their life together will be laid down for the sake of the world. The coming of the Spirit at Pentecost resulted in a bold proclamation of the gospel, the repentance of thousands, and the establishment of a common life among the believers. But the anointing of the Spirit is not just for our own religious experience; it is for the intentions of God in the world. The Spirit is the sign and seal of the church's vocation as a suffering servant. The question always before the anointed community is where and how to give its life in service of a broken world.

Community is the great assumption of the New Testament. From the calling of the disciples to the inauguration of the church at Pentecost, the gospel of the kingdom drives the believers to community. The new order becomes real in the context of a shared life. Throughout the book of Acts and in the epistles, the church is presented as a community. The community life of the first Christians attracted many to their fellowship.

The preaching of the gospel is intended to create a new family in which those alienated from one another are now made brothers and sisters in Jesus Christ. "There is neither Jew nor Greek, there is neither slave nor free, there is neither male nor female; for you are all one in Jesus Christ" (Gal. 3:28). The existence of the church itself, that inclusive community that knows no human boundaries, becomes a part of the good news.

When we understand that community is the form of the church's life in the New Testament, the letters of Paul take on a clearer meaning than ever before. Reading them in the context of Christian community illuminates their message. That is not surprising when we realize their original purpose as pastoral letters to new communities. Paul was a theologian of community. His central ministry was the apostolic task of forming and nurturing new communities. Paul hoped his preaching would create new communities; he often stayed with these embryonic fellowships until they were on a solid basis; and he corresponded with them for years afterward, offering sound teaching and practical assistance in working out their life together. The love of Paul for these little growing fellowships is evident throughout his writing. The preaching of the gospel produced neither a new school of thought nor a new political party; the legacy of Jesus was a new people sharing a new life together. Different from an institution or an organization, the church would have the style and the feel of a family, an extended family created not by blood but by the Spirit.

Paul presents us with the cosmic sweep and meaning of community. In Ephesians he says "the mystery of Christ ... hidden for ages" is now being revealed. The mystery is that Jews and Gentiles are to be united in one body in Jesus Christ. The dividing wall has been broken, the former hostility has come to an end, and a new humanity has been created in Christ Jesus. In Ephesians 2:13–22, Paul describes this new community in a way that reveals the power of God in Jesus Christ. But that is not all. Paul goes on to describe the community of believers in terms of God's plan for the world. The unity and reconciled fellowship of the new community is "to make all men see what is the plan of the mystery hidden for ages in God who created all things—that through the church the manifold wisdom of God might be made known to the principalities and powers in the heavenly places" (Eph. 3:9–10).

The testimony of such a community is living proof that the oppressive and divisive facts of the world system need no longer hold sway and determine the course of men and women. The whole of God's creation will someday be brought into community in Christ (Col. 1:15–20). The community of the church is the beginning of that great reconciliation, the sign and firstfruit of God's cosmic purposes in Christ. The church is an integral part of God's plan to reconcile all things. Through its ministry of reconciliation, the Christian community becomes both an instrument and a foretaste of God's purposes for the world. Paul describes the power of this reconciling force in his second letter to the Corinthians (5:16–20), a passage we discussed earlier.

Therefore, those who would limit Jesus to the saving of souls and those who see him merely as introducing new ethical principles are both wrong. The purpose of God in Christ is neither simply to redeem individuals nor merely to teach the world some new thoughts. God's purpose in Christ is to establish a new community that points to the plan of God for the world. Forming community has been the social strategy of the Spirit since Pentecost.[1] Community is the basis of all Christian living. It is both the lifestyle and the vocation of the church. The living witness of the Christian community is intended both to demonstrate and to anticipate the future of the world that has arrived in the person of Jesus Christ.

Conversion, then, has everything to do with community. A good test of any theology of conversion is the kind of community it creates. In the biblical descriptions, conversion is from one community to another, or from no community to community. Especially in an age of individualism and personal isolation, community becomes central to any idea of conversion. Evangelism can no longer mean simply taking people out of the world, running them through a process of conversion, and then placing them back in the same world and somehow expecting them to survive. If conversion is the translation of persons from one world to another, from one community to another, then conversion to Christ requires a new environment in which it is more possible to live a Christian life.

Nor is community simply a collection of the already converted. Community is the place where we lay ourselves open to genuine conversion. It is the corporate environment that preserves and nurtures the ongoing process of conversion. Once we have set our feet on a new road, community is what helps us along the way. In community we begin to unlearn the old patterns and to learn what the kingdom is all about. In relationship to one another, we understand more deeply the message of the gospel, and our relationships reinforce our ability to be faithful to it. The community of faith enables us to resist the pressures of our culture and to genuinely proclaim something new in its midst. Community is never withdrawn from the world, because its biblical purpose is to make Jesus Christ visible in the world.

Community is the arena in which the struggle for a faithful church will first take place. Community, therefore, does not exist for itself nor as an alternative church. Christian community is for the church. It is the battleground of the movement from captivity to renewal, from conformity to transformation. Community, then, is a living sacrament for the church. The historical issues confronting us will be first joined in communities of faith. Community can be the demonstration and the incarnation of a new word of renewal preached to the churches. As a result, community will be a place of struggle, conflict, pain, and anguish as we wage the battle with the false values around us and within us. It is where our personal and corporate sin is first revealed.

But community can also be a place of new freedom, of deep healing, of great love and joy as the power of conversion is experienced. Community helps us to grow, and it helps us to convert. We are enabled to turn from our cultural myths and illusions, and we are pointed toward the reality of the kingdom of God. Community is a place to grow in truth, wholeness, and holiness. The only way to propagate a message is to live it. That is why there can be no conversion without community. Community makes conversion historically visible.

The principal cause of the church's accommodation to the values and spirit of our age is the fragmentation of our common life. We are easy prey, because we are rootless and confused. In many places in the world today, the church suffers from brutal persecution. But in the United States our chief enemy is not persecution. It is seduction. We are a people seduced by a way of thinking, a way of living, that is irreconcilable with the lordship of Christ.

When I visit with Christians in local churches, I sometimes ask them, "What is the most important social reality of your life? What place, what group of people do you feel most dependent upon for your survival?" Very seldom have people responded by pointing to their local church, their community of faith. Instead, their answer is their workplace or some other economic, educational, or political institution. People usually name something associated with economic livelihood, personal advancement, or social influence.

If in fact most Christians are more rooted in the principalities and powers of this world than they are in the local community of faith, it is no wonder that the church is in trouble. Clearly, the social reality in which we feel most rooted will be the one that will most determine our values, our priorities, and the way we live. It is not enough to talk of Christian fellowship while our security is based elsewhere. We will continue to conform to the values and institutions of our society as long as our security is grounded in them.

We need to know where our securities lie because they can and will be used against us, even as we begin to enter deeply into Christ's community. Our securities will be used to intimidate and control us, to rob us of our freedom in Christ. Just when we begin to respond to God's calling, we often move out into insecure places; this is when the powers of this world reach out, hook us, and reel us back into their circle of control. Where they hook us is at the point of our deepest insecurities. We are attacked in the places where we are most vulnerable and most easily controlled by the rewards, threats, and punishments of the system.

The only alternative is to create a faith community that generates a faith strong enough to enable us to survive as Christians. This strength is important at two points: in helping us to disengage from the securities of the old order and, at the same time, in empowering us to be actively engaged in the world as witnesses to the new order in Christ. Community is the place where the healing of our own lives becomes the foundation for the healing of the nations. The making of community is finally the only thing strong enough to resist the power of the system and to provide an adequate spiritual foundation for better and more human ways to live.

At a minimum, the church should be known as the kind of community that makes it more possible, not less possible, to follow Jesus. But this is not always the case in today's churches. A New Testament scholar once told me, "I have a hard time teaching my subject because, when I get to the idea of the community of faith, there is little I can point to today to show my students what it means. There's no problem, of course, describing what it meant back then. But I don't know how I can help them to understand when there are few examples I can point to now." The statement is enough to make one more than sad; it should make one angry—angry at the control the system has over the church's life today. There is reason to be angry about a system that crushes poor people and defines whole populations as expendable in war. There is justification for anger at the ways the system has crippled and co-opted the faith of the churches. But the target of that anger is misdirected if it is aimed at the people who are trapped.

Jesus was full of anger when he entered the temple and thoroughly disrupted the business of the day. But he was an-

grier at what the people were doing than at the people themselves. He was angry at how the economic system was making
a sacrilege of religion in the temple. Are we angry at the way
the economic system has made a sacrilege of faith in the local
church? The deeper our identification with the church, the
angrier we ought to be. However, we must get beyond an ad-
versary relationship with the people. The powers that be must
take great pleasure in the way we constantly fight each other.
We have to get beyond the spirit of accusation.

Those who have experience in pastoral counseling know
what it means to be involved with persons who have lost all
vision for their lives. The pastor's task is to hold before those
people the vision of their wholeness and healing until they
can grasp it and claim it as their own. So it is with the church.
The church has lost any vision for its life together. Without
that vision, the people are perishing. The vocation of those
committed to rebuilding the church must therefore be to hold
forth the vision of a renewed people, the vision of healing and
wholeness. This is the role of conversion. Conversion begins
by calling the church to repentance, by calling God's people
back to a new understanding of who they are and to whom
they belong as God's people. To hate the church for its failures
is like a pastor hating the person who needs healing. Only
those who have come to feel a genuine love for the church
will be able to confront it with its own faithlessness and call it
back to its true vocation.

I am often asked if I believe that real Christian community
is possible in the established denominational churches. The
question reminds me of what Mark Twain once said when
asked if he believed in infant baptism. Twain replied, "Believe
in it? Hell, I've seen it!" That's my answer to the question
about the established churches. I've seen Christian community

take root and grow in the churches. Many people in the institutional churches admire noble ventures into community but consider them irrelevant to their own or other local churches. The church has a tendency to put radical communities up on some inspirational pedestal, something to point at but not to imitate. Catholic Worker founder Dorothy Day once spoke to the danger of being admired into irrelevance. She said, "Don't call me a saint. I don't want to be written off that easily."

There was a time when I almost regarded alternative communities as the "real church" and the institutional churches as the "apostate church." I don't do that anymore, for a number of reasons. First, I have discovered that most of the problems that exist in the church also exist in my community and in my own life. Sojourners community was always a microcosm of the problems that are faced in the larger church's life. We are all full of this world, full of this culture, and we are in a slow process of being converted. Increasingly, we were able to identify not just with the strength of the church but also with its many weak and broken places.

Second, most of the churches today began their tradition by deciding to split off from the "apostate" church and become the "real" church. The cycle goes on and on, creating more versions of the apostate church. Personally, I am not of a mind to create new denominations and new divisions in the church's life. I am more anxious to speak of a new vision for the church, realizing that this new vision is in fact two thousand years old and, in most cases, is a vision that exists somewhere in the theological tradition of most churches. The task is to point these churches to the seeds of renewal in the Bible and in their own traditions. In most of the churches where renewal is taking place, there is a fresh emphasis on Bible study. The Bible is coming alive again as congregations reflect on their own experiences and relate the biblical word to their present historical situation. When the Bible is used simply to affirm and sanctify the present order of things, it is emptied of its power. But studying the Bible in a way that calls present realities into question will uncover its tremendous power to heal us and to change the world. The biblical word creates in us both the need for conversion and the hope of conversion.

Many of us tend to underestimate the hunger in the churches for something different, some new vision and focus and power. I often sense in the churches an underlying uneasiness about feeling so at home in this culture. There are people scattered throughout the churches who sense that their commitment to Jesus Christ ought to mean more than it does. There is a desire and a fragile hope for something new. In spite of all we have said about the American captivity of the churches, an integrity of faith remains in the church's life. In most churches I have visited, a small flame flickers that invites rekindling. As a result, I'm not ready to give up on the churches, and I'm certainly not willing to give up on the gospel.

Finding a new way of life for the churches, a new shape for our corporate existence, is finally a question of pastoral leadership. If the fullness of the gospel is to be preached and lived in the churches, the responsibility for this begins with our pastors. The pastoral vocation is to testify to what we know of the gospel. We cannot suppress the gospel in response to some mistaken notion of "being sensitive" to the needs of our church members. When for any reason we fail to preach and to live the gospel in its wholeness, we fail not only our prophetic calling but our pastoral calling. To fail to lift up Jesus in the midst of our congregations and in the midst of our history is a failure of love—a failure to love one another and to love the church. Ultimately, of course, it is a failure to love God deeply and to trust God's deep love for us.

The prophets of the Bible spoke the hard word to a captive people. They were angry at the people's infidelity. They spoke the Word from the Lord, and often the people didn't want to hear it. But when the prophets spoke, they spoke with broken hearts, because they knew the people, loved them, identified with them, and held them as their own. Undergirding the prophetic rebukes was the vision for the faithful life of the people of God. The prophets didn't speak from arrogance, pride, bitterness, or despair. They spoke from love and hope for the people. Hope is ultimately rooted in love. So our hope for the church must be rooted in our love for the church. And love is the great enemy of fear. So our love for the church must overcome our fear of being the church, our hesitancy to preach God's word and to give visible demonstration of God's vision for the community of faith.

In Christian community, we are still learning what it means to love. God keeps teaching us much, softening our hearts and expanding our capacity to love one another and the whole of God's creation. The process is always one of conversion. The "turning to" part of conversion enables God's love to deepen among us in some exciting ways. But the "turning from" part of conversion has never been easy. We learn that all our models and schemes for community have to die before God's creative work among us can begin. Our plans and pride over what we can build with our own strength and resources have to be shattered before the Spirit has any room to work. And we have to learn that the necessary building materials of Christian community include two characteristics of love: forgiveness and a humble spirit. Being human, we cannot avoid conflict and hurting one another.

We must realize that we are utterly dependent on God's forgiveness in our corporate life. Learning to forgive one another, and to know our own need for forgiveness, are lessons that test the survival of the community. We also have to get over any notion of being perfect people building the perfect community, which can then take on all the big issues of the church and the world. The big issues overwhelm us, because we forget to tend to the simplest things, like learning to love and serve one another in our imperfection. The lesson here is a basic one: The church will never discover what it means to lay down its life for the world until its members begin to lay down their lives for one another. An authentic public witness requires an authentic community existence. The love, care, justice, and peace we desire in the world must also be practiced among ourselves.

The words of Scripture began to take on new meaning for us. Jesus said, "A new commandment I give to you, that you love one another; even as I have loved you, that you also love one another. By this all will know that you are my disciples, if you have love for one another" (John 13:34–35). Jesus tells us to love each other, not simply *because* he loves us, but also in the same way that he loves us. We are to extend to one another the very same love that God has extended to us in Christ. We are told to love as we have been loved, to forgive as we have been forgiven, to share as we have been shared with, to sacrifice as we have been sacrificed for, to reconcile as we have been reconciled, and to make peace as peace has been made with us.

Conversion means a radical reorientation in terms of personal needs and ideas of personal fulfillment. When we enter community we bring with us an emptiness that seeks filling, but we also bring clear notions of what we think might fill that emptiness. We know our own needs best of all, and we are fairly sure about how they can be met. All of us, sooner or later, have to put aside the primacy of our own needs; we have to relinquish our narrow expectations of self-fulfillment and our agendas for self-assertion. Conversion is ultimately dying to self and becoming part of something that is larger than any of us. Community is the environment that can enable that conversion, and community is the fruit of that conversion. Our perspective changes from "what can the community do for me?" to "what can I do to best serve the community?"[2] The ramifications of this conversion are profound. The change affects us spiritually in terms of our identities, politically in terms of our loyalties, economically in terms of our securities, socially in terms of our commitments, and personally in terms of our vocations. Through it all, the most profound change is finally the most simple: discovering the meaning of love.

> If I speak in the tongues of men and of angels, but have not love, I am a noisy gong or a clanging cymbal. And if I have prophetic powers, and understand all mysteries and all knowledge, and if I have all faith, so as to remove mountains, but have not love, I am nothing. If I give away all I have, and if I deliver my body to be burned, but have not love, I gain nothing. (1 Cor. 13:1–3)

Paul's words are a confirmation of the fact that love is the strongest, truest, most powerful, and most revolutionary gift. The passage reminds us that the Christian life begins with love, and it also ends with love. As Christians there is nothing we can do, nothing we can become, that is prior to love. As we come to faith, we begin to love. Love binds us to God and to God's people, and it frees us to minister to one another and to the world.

We can love because we are loved. God's love for us, made flesh in Jesus Christ, becomes the basis for our love. The love of God for men and women precedes the Christian life. Our love is simply an answer to God's love for us. Love is the center, the essence of Christian life and community.

Love changes us. The greatest conversion within us is the deepening of love among us. People made hard and cynical after years of frustration and discouragement can become gentle. Those who once knew fear now know trust. People bound to the materialism of the world discover the freedom of simplicity. Men trained well in the ways of power and control come to terms with their own vulnerability. Women held back in restrictive roles assume strong leadership in the community. Whites reared in cultures of racism become hard workers alongside low-income black tenants. Many persons schooled in the ways of competing and winning learn how to devote their energies to active peacemaking. The strong are put in touch with their weakness, and the weak find strength within themselves that they didn't know they had.

The environment of love always opens up new gifts. The power of God's love is active in deepening gifts already evident and channeling them in new directions. Gifts only partly known by certain people are nurtured and expanded. Other gifts are called forth from people who never knew they had them. Some of the elicited gifts are ones that people buried in the past because they didn't seem to count in the world. The freedom that comes from being loved enables each of us to discover many new things: things about God, about each other, and about ourselves. Our capacities to receive love freely are stretched, increasing our abilities to love others more freely. In the process we learn more about God's love. All of this deepens our conversion, because to be converted is to know the fullness of God's love and to live it out.

We have nothing more to share with the world than what we are sharing with each other. We can effect no change in the ways of the world unless we ourselves are being converted from those ways. On the other hand, seeing how people's lives are being healed and changed in community gives us hope for the world. If God can change us, maybe God can indeed change the world. The very things that the world so desperately needs are things we begin to see and experience in our own community. As love increases, so does hope.

We also learn important things about reconciliation. All communities are made up of diverse people. We come with different personalities, backgrounds, temperaments, and political and spiritual traditions. To be reconciled one to another must be a primary community. Otherwise, there would be no community. A commitment to reconciliation requires doing whatever is necessary to remain in fellowship with one's brothers and sisters. Nothing is hidden in life together; all of our hurts, fears, sins, joys, struggles, and hopes are exposed. When this creates difficult relationships, and we avoid the necessary reconciliation, we have failed. The implications are large, for we cannot hope to reconcile in the world that which we cannot reconcile in our own lives. If our own differences overcome us, how can we be peacemakers in a world full of differences, conflicts, and violence? We learn much about reconciliation, how it happens and what it takes from us in terms of caring and investment in one another. And what we learn can teach us much about the meaning of reconciliation in the world.

The ability of people to move to a new place tomorrow

depends on the love and acceptance they feel today. In community it is hard not to know everything that is wrong with each other, including the sins and mistakes of the past. With this awareness comes a choice: We can complain and judge the other person, or we can love him or her the way Jesus loved the woman at the well. The only thing greater than our awareness of each other's sins is the awareness of God's love for us and God's desire to see us healed and made whole. This is the kind of love that deepens conversion in community. Conversion begins with turning, and forgiveness invites that turning. For most people, the experience of God's forgiveness occurs most directly through the forgiveness of their brothers and sisters. Only out of that forgiveness are people enabled to move from their past into God's future for their lives. Communities that develop a habit of forgiving and loving also tend to develop an overflowing reservoir of forgiveness and love, something they cannot help but offer to a world full of pain and suffering.

The principal lesson of community is a principal lesson of the kingdom—namely, that God breaks in at the weak places. God's Spirit is active in the most unlikely places—the poor, broken, and humble places. The power of God is most realized at the point of our vulnerability, our risk-taking, and our letting go. To be vulnerable means to be available to the power of God's love. Community brings us to the point where God's love can break in. Most people seem to think that community is for the weak, for those who are broken by or just can't make it in the "real world." The assumption, I guess, is that community is for those who can't support themselves and need the support of a group. I cannot deny that people in their weakness need community. But I often think that those people in our society who consider themselves strong and whole, by the world's standards, need community even more. These are the persons who can experience the deepest conversion in community. Here, for the first time, their weak places are acknowledged and accepted. In community they need not hide nor fear their insecure places. No longer do they have to posture and play to their strengths; they can be accepted and indeed loved in the wholeness that is strength and weakness.

All of us need the assurance that we are loved, the confirmation that we are "all right." Our identity and security depend on that love and confirmation. Only then do we feel safe and secure, fully capable of loving and being loved. Christian community is the environment that provides confirmation, comfort, and challenge. We no longer need the authority of a system to affirm and authenticate us, because we now recognize and worship a higher authority. When our source of security becomes God alone, we can safely act independently of the systems around us. The love made possible through Christian community can provide the necessary inner authority we need to act more faithfully in the world.

The body of Christ is a rich biblical metaphor. It means to be so close to others that you feel as if you were one body. When one part suffers, all suffer. When one part rejoices, all rejoice. We share a common life, a common ministry, a common calling, a common worship, and a common destiny.

The crucial relationship between vision and nurture are central to the experience of community. Both are key to conversion. Without nurture, a community will soon exhaust itself in pursuit of the vision. Without vision, a community will become stuck in self-preoccupation and will travel in circles. With only vision, a community soon loses any real quality of love. With only nurture, the community forgets what its love is for. Vision without nurture can be oppressive and destructive; it will place an overwhelming burden on people. Unless people are being nurtured in the vision around which their life is called together, there will be no community. Similarly, without a prophetic voice challenging God's people to lay down their lives in the world, pastoral nurture can easily degenerate into self-serving group welfare or inward and unbiblical withdrawal.

One of our greatest struggles is to find a unity and integration between the prophetic and the pastoral life. The separation between the two is perhaps one of the most pervasive divisions in the church today. To experience a wholeness and rhythm between these two styles of life is certainly a biblical necessity as well as an urgent priority for the fragmented times in which we live.

The relationship between the pastoral and the prophetic imperatives converges in what we believe is the basic theological question of our time: the shape of the church. We have come to that conviction out of our pilgrimage as a people and out of our struggle to comprehend the meaning of biblical faith and the prophetic character of the church's life and witness in the world. We feel that to talk of identification with the poor, to talk of peacemaking, to talk of forging a new lifestyle, and to talk of Christ's love becoming incarnate in the midst of the world's pain—all of this is to talk about the shape of the church. Similarly, to speak of the renewal of worship, to speak of personal healing, to speak of recovering the pastoral life and ministry, and to speak of discipling and evangelism—all this is also to speak about the shape of the church. The deepest issues for Christians return us to the underlying question, which is the shape of the church's life in the world.

Building the body of Christ is not one of many issues to which we are committed, as it once was; it is the basis for all that we do and all that we are. It is the environment in which and out of which we are called to live and minister in a world of pain and peril. We can confront the violence of our age only with the very life of Christ among us. And we can confront the economic imperatives that lead to war only by making visible the love and simplicity that was Jesus's way of life. It is not us but his life among us that disarms the principalities and the powers (Col. 2:15).

The call to peacemakers is, first of all, to live at peace with one another. The quality of our community life is crucial:

Does it nurture us in the way of peace, or does it distract us?

Does it free us from bondage to material goods and security?

Is our Christian fellowship healing us of our hate, fear, selfishness, desire for power?

Does our experience in the local church root out those things that are foundational to the system of injustice and violence?

Christian community can put us in touch with the fear and the violence still within and among us. It can help us to understand the grip of those forces in the world; we can learn the things that make for peace in our life together and offer them to the world.

Most of us in local congregations are not yet free enough to be peacemakers. Our faith is not operational at a level deep enough to break the hold of the habits and systems that lead to violence. Peacemaking ultimately requires a maturity in our spiritual life that most people have yet to experience. Christian community can help to develop that kind of maturity. To the extent that there is a strong pastoral environment, community can be a training school for peacemaking. Community, like nothing else, puts us in touch with our own sin and propensity toward violence. Community with brothers and sisters also becomes the place of God's healing. In a healthy pastoral environment, we learn the ways of resolving conflict, overcoming prejudice, healing our fear, and being reconciled with our enemies. Those who have experienced the things that make for peace are desperately needed in a world of violence.

At the same time, many people who know the healing life of a pastoral community have yet to see the implications of their community life for the world. A community life based on forgiveness and reconciliation extends naturally into active peacemaking in the world. Those who have learned to love their enemies in community are the best persons to offer their experience in the wider political context. Christians should be people who generate trust and reconciliation wherever they go. Paul's list of the fruits of the Spirit reads like a prescription for survival: love, joy, peace, patience, kindness, goodness, faithfulness, gentleness, and self-control (Gal. 5:22–23). Community is the crucible for our conversion to these virtues; it is where we take our first steps in learning to walk by the Spirit.

The recovery of the gospel for these times will come from knowing these fruits of the Spirit in our life together and offering them in the midst of economic injustice and violent threats. We must resist the temptation, however, to over-spiritualize the gifts of the Spirit. We are called not just to have a right spiritual attitude; we are called to live in the Spirit in concrete and specific ways that are an unmistakable alternative to the ways of the world.

It is vital that groups of people say, "Wait, we do have a choice; there is another way. We are people just like you—men, women, and children no different from you—and we are able to live quite well, rather simply, and very free from the materialist impulses of the nation. We have learned another way to live that is not only economically viable but healthier and more creative. We have less of the world's abundance, but life is more abundant than before. We are happier and freer, and we'd like you to be too." The presence of such people would be an example and an invitation. It would not just tell society that "there ought to be another way." It would show that another way already exists and that all are welcome to join and to share in the riches of the new order. And the presence of such people would be highly relevant to the political economy of a nation that ultimately trusts violence to solve its problems.

The early church was built on small groups of people who came together to support one another in a whole new way of life. These primitive communities were the evidence of a new order. Today we need small bands of people who take the gospel at face value, who realize what God is doing in our time, and who are living evidence of the reality of conversion. People need to see practical demonstrations of another way to live and to understand its larger relevance for justice on a global scale. To see the deepening reality of another person's conversion is to believe that even we can be so changed. Simplicity, sharing, and living in a way that embodies love and makes justice more possible is normative for all of God's people. Our preaching has to make that clear. Our need is for prophetic preaching and practical demonstration.

One of the effects of such evangelism would be the spread of social deviance. The priorities of God's kingdom are at such variance with the ruling assumptions and structures of our day that the simple proclamation of the kingdom could undermine the economic system and present the risk of being charged with political treason. The very idea sends tremors through a church that has worked hard to achieve majority status in this land. But the New Testament church was self-consciously a minority in its cultural context, and the church in the United States might dare to recover a similar minority status. When the church dares to be the church, it becomes a self-conscious alternative to the mainstream culture. To the extent the church allows the gospel to intrude into its life, the church regains its distinct identity over and against the values of the world.

I think we are at this moment right on the edge between captivity and revival, between conformity and renewal. We may be about to regain some gospel integrity in the church's life. Even though none of us has all the answers, there are people beginning to ask the right questions about the shape of the church today. Most often at the local parish level, questions about faithfulness in our historical situation are being asked. People are seeking a new place for faith and a new experience of one another at the congregational level. There are hopeful signs of new life in the church.

Several things are true of a growing number of Christians in this country. They are learning that their worth and identity as human beings does not depend on their consumption and possession of things. They are Christians who no longer believe the central lie of the economic system. They neither shun nor ignore material needs, but meet them simply as they care for one another in communities of faith. There is no longer an ultimate financial incentive in their lives; economic success is no longer the goal. Material goods now have only instrumental value.

A change has come about in the way they tend to look at the world. Social questions, political decisions, and newspaper headlines are now viewed from the vantage point of how they affect poor people. For most of them, that is an entirely new starting point, a whole new perspective for how to think and act. Now, their first impulse is to ask, "How does this affect the poor?"

These are most significant and hopeful things, because they demonstrate a new belief in God as well as a new disbelief in the two most basic assumptions of the present system. The fundamental problem today economically is that people believe the myth that economic gain is the key to happiness. The fundamental problem militarily is the myth that security comes from more and more weapons. Those myths are dying among many Christians. And, to the extent they are no longer believed, the system has lost its legitimacy for them. The success of the U.S. economic system depends on people identifying themselves principally as consumers. If even a minority begins to define itself differently, that system is threatened. Similarly, when people fear the military arsenals more than they fear the things the arsenals claim to protect us from, the military system loses its credibility. People begin to believe in peace more than war in resolving conflicts in a more effective way.

The power of any system is finally not based on its wealth, technology, or military hardware. Ultimately, its power is derived from the level of spiritual authority it has in people's lives. In other words, a system has power only to the extent that people believe in it. When people no longer believe the system is ultimate and permanent, when the myths begin to die, the hope of change emerges.

Many of us were born and bred to be the managers and beneficiaries of the present system. Now we no longer believe in its most basic assumptions. That is social change. It is a new social vision in the making, one that is based on reclaiming all of the good news of Jesus Christ. This is exactly what the society and the world most need today: a new social vision. A new perspective is not enough. A whole new vision is needed, and the church has the possibility of offering such a vision out of its own life together. A church living by biblical economics would be made up of ordinary people who broke with the economic givens of the society. Simplicity would replace accumulation, and sharing would replace competition. Those people would be concrete proof that it is possible to live a different way. Their voice would be a clear and credible defense of the poor and an attack on the oppressive arrangements of wealth and power. The church's very existence would show forth the promise of new social possibilities and new economic arrangements.

We can also envision congregations of Christians around the world who sense the urgency of peace and justice. People drawn from all of the world's warring factions, reconciled in Christ, would be particularly well situated to show the way to peace, to help fearful nations learn less destructive ways to resolve conflicts. In communities of faith, where the war system has been renounced as spiritually idolatrous and politically suicidal, concrete initiatives could emerge for beating swords into plowshares. This could be the new shape for the church. And this could be a new social vision for the world's anxious and suffering people.

If there is hindsight for our time, it will show this period in history to be a time of transition, a time of change from one era to another. The assumptions, values, and structures that gave rise to the present age are unraveling; they are no longer adequate. But the new assumptions, values, and structures have yet to take concrete shape. It is a time of transition in which people are understandably nervous about their lives, their futures, their world. Many people are tending to look at tomorrow as that time when today's problems only get worse.

The signs are all about us, pointing to the great need for a new social vision. When confusion and uncertainty abound, the future belongs to those who can see it and begin to live it. We need a new understanding of how people can relate to one another spiritually, socially, economically, and politically. Our society is crumbling for want of a vision that has the capacity to change personal lives and to generate new social and institutional patterns.

That new social vision will most likely arise from religious roots. In U.S. history, major social transformations have most often grown out of religious revival and spiritual awakening. The abolitionism of the nineteenth-century revivalists is a prime example. The renewal of faith, more than the spread of ideology, has been the catalyst for change. The changes now necessary have to do with our most basic values and beliefs—conversion. The needed changes involve questions of ultimate reality and ultimate authority in people's lives. In short, the needed changes extend to our spiritual foundations.

Social disintegration should not, however, be viewed simply with despair. It can be, in fact, a sign of hope. We are in danger when disintegration leads only to despair, for despair breeds passivity and becomes yet another victory for the system. Biblical hope comes from having a vision of the future that enables us to live even now in its promise. It means bringing the future into the present with power and authority. Hope in something new and more promising is always the greatest spark for change. Without that hope we are controlled by present realities, wandering between passivity and despair.

We are witnessing a battle for the minds and hearts of people. Some would try to channel collective insecurity into a rigid, ideological agenda that reinforces the worst values and structures of the present system. They would prefer that we didn't go to our spiritual foundations seeking new hope. Others would like us to feel that, while there are some problems in the world, we should remain optimistic.

The crucial movement for people of faith is from optimism through despair into hope. It is the ability to see through the false optimism of the times, to see the reality of despair in the world, and to see the hope that ultimately comes from God's loving purposes for all creation.

In the midst of a hopeless time, that movement will emerge in the living out of a new biblical vision, and it will offer hope. The situation need not improve to give rise to hope. All that is needed is a belief in the possibility of an alternative. Jesus has shown us the alternative, and he asks us to believe. He knows our loneliness and despair, and he prays for our fellowship and joy:

> But now I am coming to thee; and these things I speak in the world, that they may have my joy fulfilled in themselves. I have given them thy word; and the world has hated them because they are not of the world, even as I am not of the world. I do not pray that thou shouldst take them out of the world, but that thou shouldst keep

them from the evil one. They are not of the world, even as I am not of the world. Sanctify them in the truth; thy word is truth. As thou didst send me into the world, so I have sent them into the world. And for their sake I consecrate myself, that they also may be consecrated in truth. I do not pray for these only, but also for those who are to believe in me through their word, that they may all be one; even as thou, Father, art in me, and I in thee, that they also may be in us, so that the world may believe that thou hast sent me. The glory which thou hast given me I have given to them, that they may be one even as we are one, I in them and thou in me, that they become perfectly one, so that the world may know that thou hast sent me and hast loved them even as thou hast loved me. (John 17:13–23)

Jesus prayed for his disciples and for all those who would come after them. He knew that they would be in conflict with the world. He knew this because his own life, his words, his kingdom were in such conflict with the world; now his life is in them, his words have been received by them, and the message of his kingdom has been entrusted to them. The world will hate them just as it hated him. The world will treat them as it treated him, because they will not be of the world. They will not belong to the world but will live as strangers in it, just as Jesus did.

And yet, for all the expectation of conflict in Jesus's prayer of intercession, it is not a prayer of despair, bitterness, or pessimism. Rather, it is a prayer of deep love, filled with hope and joy. Jesus yearned for his disciples to know and to be sustained by the same love that binds him to God. The unity that he has with God he wanted to share with his disciples, "that they may become perfectly one" and "that they may have my joy fulfilled in themselves."

The power of the Christian life is joy and hope in the face of discontinuity. The churches have never accepted this easily. Endless theologies have been constructed to ease the discontinuity, to reduce the conflict, to find some accommodation between Christ and the world, to affirm the world on its own terms, to secure a comfortable place in the world, and to find our hope there after all. The placing of false hope in the world and its power to save itself has always been and continues to be a great threat to the church.

What the church must always seek is the gracefulness of life lived in discontinuity and conflict. It is the gracefulness of living an ordinary and normal life in Christ that is so extraordinary and abnormal in the world. Partaking of the richness of that life, one that the world regards as a scandal, is the source of our joy.

Those Christians who have experienced the conflict between the gospel and the world most personally and painfully are always the ones who have known the joy of Christ most fully. The stronger the identification with Christ, the deeper the conflict, the greater the joy.

In Jesus's prayer, the nurturing of the love the disciples shared together is the key to their position of discontinuity with the world. The richness and power of their love for one another made their uncommon existence in the world not merely tolerable but renewing and life-giving.

On the eve of his crucifixion, in which his own conflict with the world reached its climax, Jesus's principal concern was for the quality of life shared by his disciples. That same concern must be our own, especially when our conflict with the world is most pronounced. It is the vitality of our love for one another that makes the radical Christian life not something to be endured but to be celebrated. The hope and joy of this celebration makes possible our resistance; it saves us from the cynicism, bitterness, and hatred that would otherwise be the consequences of a life lived in opposition to the world as it is.

Everything we mean by conversion must take human form and flesh in the place we call community. Such a life is more than the Christian's only hope. It is also the world's only hope.

CHAPTER 6

The Roots

I appeal to you therefore, brethren, by the mercies of
God, to present your bodies as a living sacrifice, holy
and acceptable to God, which is your spiritual worship.
Do not be conformed to this world but be transformed
by the renewal of your mind, that you may prove what
is the will of God, what is good and acceptable and perfect.

Romans 12:1–2

SINCE THE EARLY DAYS of Sojourners, this passage has
continually taken on new meaning and importance in our
community life. In our formative years, we saw that Paul's
words "Do not be conformed to this world" spoke directly
to the cultural captivity of the American churches. As we
grew, our concern expanded from one of simply desiring to
break free of our conformity to culture. We experienced a
deep desire for conversion, to be "transformed by the renewal
of your mind." In more recent times, we have noticed that the
context of the whole passage is worship; specifically, Paul is
concerned about true worship and false worship.

For Paul, worship means "to present your bodies," or, in
our language, "to offer your very selves." In true and genuine
worship, everything we have is to be offered to God; every
part of our lives is presented as a "living sacrifice." Worship,
then, is not limited to a ceremony, an occasion, a service, a
place, or a particular time during the week. The worship

events that punctuate our lives are intended simply to gather and focus all the parts of our lives. According to Paul, all of life, all of who we are and all that we do, is the content of our worship. Worship is the drawing together of the whole of our lives and the offering of that whole to the Lord. All of our humanity and all of our relationships—to God, to the world, and to one another—are brought into worship. We offer our brokenness and remember our wholeness. We offer our alienation and uprootedness, and we recall our rootedness in the history of God's people. To offer ourselves so totally to God is what Paul describes as "spiritual worship."

Paul's next words are about conformity to this world. In the Phillips translation, the words are "Don't let the world squeeze you into its mold." The New English version says, "Adapt yourselves no longer to the pattern of this present world." Paul is saying that conformity to the world is more than a failure of lifestyle or politics; conformity to the world is a failure of worship. The integrity and vitality of our worship is tested in our relationships to the world. Paul, knowing how often we will fail these tests, calls us to be transformed. He wants us to have our minds renewed so that we may know the will of God, that we may know what God is saying to us, that we may know "what is good and acceptable and perfect." Without this transformation, we will continue to conform to the patterns and structures of this world; without this renewal, we cannot know the will of God because our attachments to the world cloud our vision and obstruct our hearing.

We live in a time when worship seems to lack power and authority in many congregations. We would like to think that we can renew our worship simply by developing better liturgy or music; by more creatively using dance, drama, and the other arts; or by learning more effective ways to lead congregational worship. Paul wants us to look much deeper. He wants us to understand that none of these ideas will work as long as the church conforms to the systems of the world. That is where our transformation, and our renewal of worship, must begin.

Worship has become an important part of our life and work at Sojourners. In our prayer times together and in staff chapel, we are refreshed and renewed, our vision is rekindled, and we remember again who we are and what we are about. People who know some of our history are often surprised by our emphasis on worship. They suppose us to be solely concerned about the political meaning of the gospel, not about the renewal of worship. And many communities around the world that began by emphasizing renewal of worship are now finding increasing political meaning in the gospel. All of this could be surprising; none of it should be surprising.

When I was a new Christian, I read something by a monk who said, "Worship is the principal vocation of Christians." I remember disliking what he said. I was concerned about the gospel imperatives of feeding the hungry, clothing the naked, sheltering the homeless, and working for peace in a world at war. My dislike for his statement, in hindsight, was based on an inadequate understanding of worship. To put it bluntly, my idea of worship was extremely provincial.

I was the victim of a trap that continues to confound and paralyze much of the church today. It's the trap of dividing the church between those who regard the gospel as principally spiritual and those who see the gospel as primarily political. We are either concerned about worship and our pastoral life or we are concerned about the social imperatives of the gospel. This schism has torn the church in half and impoverished everyone on both sides of the dividing line. Our basic premise must be that the gospel is made of one cloth woven of the same fabric. That is to say, what God is doing within us and among us and what God is doing in the world are finally the same thing.

Since the earliest days of the church's life, there has always been an integral relationship between worship and politics. Worship and politics both raise the same questions: Whom do we love most? Where is our security finally rooted? To whom or to what are we most loyal? What finally is our deepest identity? All of these are worship questions; they are also all political questions. Each of the questions raises the issue of fidelity. Our answers reveal where we have our historical, spiritual roots.

In the days of the early church, the Christians were often accused of being atheists. Their love for the Lord, the fervor of their worship, and their fidelity to Jesus Christ were so clear, public, and unmistakable that they were charged with being atheists in regard to the false gods that ruled the Roman world. The central affirmation of the early Christians was "Jesus is Lord." They knew and the authorities knew what that meant. Their worship was understood politically by the authorities, and they were treated accordingly. If Jesus is Lord, then Caesar is not. If Jesus is Lord, then neither mammon nor the nation is the object of worship. If Jesus is Lord, then his disciples have no other lords. In short, worship expresses to whom we belong. A fair question for the church today would be: Is our worship of Jesus as Lord so utterly clear and publicly visible that we are accused of atheism in regard to the false gods of our culture?

Worship calls us back to the roots of our identity as the people of God. "Once you were no people, but now you are God's people" (1 Pet. 2:10). Worship is knowing all of our humanity in the presence of God. The people of God are known in the world for the same things for which their God is known. God's people should care about the same things that God cares for. Our purposes and priorities are the same. We love the same things, hate the same things, take joy in the same things, and hurt over the same things that God does. We are a people called to bear the very likeness of Christ in our life together and to reflect his character in the world. When we understand this, then identifying with the poor is neither first nor last a social concern; it is, from beginning to end, a statement of worship. Similarly, becoming known as peacemakers in the world is not simply a political stance; it is through and through a reflection of our worship. We worship a God whose purposes in history are clear. We worship a Lord who embodied those purposes in history. As God's people, as followers of Jesus, we not only come to know these purposes in our worship; we also make visible God's purposes in the world, identifying with them as a matter of worship.

When we worship in this way, we make a statement about where our security ultimately rests. We stand before the world free of its securities. Nothing is more threatening to a system than people who are free of its control, free of its rewards and punishments. Pilate said to Jesus, "Do you know that I have power to release you, and power to crucify you?" And Jesus answered him, "You would have no power over me unless it had been given you from above" (John 19:10b, 11a). Pilate, the symbol of Roman power, expected at least a modicum of respectful worship from all the people. So he reminded Jesus that he held complete power over him, the power of life and death. But Jesus worshiped a higher authority; he stood free of anything Pilate could do either to win him over or to defeat him. Jesus's security was rooted in his communion with

God; that is why he was a political threat to Pilate. Jesus demonstrated a new kind of security and freedom in his life, and he offers that same security and freedom to us. Jesus knew to whom he belonged, and he invites us to remember to whom we belong. He points us toward and invites us into worship.

In the German town of Flossenburg is a tablet with the following inscription: "Dietrich Bonhoeffer, a witness of Jesus Christ among his brethren. Born February 4, 1906, in Breslau. Died April 9, 1945, in Flossenburg." Bonhoeffer was executed in a Nazi prison camp. Much can be said about his theological brilliance, his political resistance, his personal courage, his warm pastoral heart. But none of this is memorialized on the tablet. He is described simply as a witness of Jesus Christ. That is all that needs to be said; nothing else matters. Bonhoeffer's life made Jesus Christ visible in this world, amid forces and powers that sought to obscure his presence.

There can be no higher calling for any of us. Discerning where and how Christ is at work in us and in the world is our vocation at its best. Our witness is strongest when we have known the presence of Christ among us and have shared that life and that love in a suffering, warring world. Our witness is weakest when our captivities to the ways of the world have obscured the presence of Christ among us. At those times we have nothing new to make visible in the world; we simply mirror the brokenness that is all around us. We begin to distrust each other and God's purposes for us. We forget how and what to worship. Specifically, we begin to chase after our own idols, forgetting our roots in God.

Every idol is defeated by the worship of God. All false worship is exposed in the light of true worship. In the early days of the Sojourners community, we used to think that all we had to do was simply to unhook ourselves from the captivities of American civil religion. We were deeply conscious of the leading cultural idolatries and of how they could be dethroned by genuine worship. While we were busy at the front door confronting the well-established cultural deities, though, some other idols snuck in the back. Except for the tender and transforming power of worship among us, we might have exchanged our allegiance to the gods of the state and the idolatries of the economic system for a new allegiance to the gods and idols of the opposition. We were tempted to make an idol out of our simple lifestyle. Our identification with the poor threatened to become an idol. We were tempted to idolatry in our actions of public protest. Our principle of nonviolence became an inviting idol. We were tempted to make an idol out of our prophetic vocation. And through it all, there was the danger of making idols of ourselves, of taking pride in our role as "radical Christians." Having exposed our idols in the larger culture, we came face-to-face with other idols closer to home. Being closer to home, they were harder to recognize.

In worship, we are reminded that the motive for living simply is that we might love both God and the poor more freely. It is not a badge of righteousness, nor something used to judge others, nor a duty born out of obligation or guilt. It is not a simple lifestyle that justifies us. Rather, it is God's grace that enables us to live more simply, more freely, and more joyously.

Worship calls us back to God's identification with the poor and Christ's presence among the lowly and afflicted. We are not to romanticize the poor amidst visions of our own downward mobility. Nor are we to exploit the suffering of the poor, building personal careers on their oppression. We are to identify with the poor, not to save ourselves, but so that we might better identify with Christ. He is already among the suffering and the forgotten, and he invites us to join him there.

In worship, we are reminded that all our actions of public protest must be rooted in the power of love and truth. The motivation for direct action in the public arena must be to open people's eyes and to soften their hearts, not to prove our righteousness and their wrong-headedness. Our actions do not have the power to save us; they can have the power to make the truth known. We are called to admit our complicity in the evil we protest; we are called to public witness marked by a spirit of genuine repentance and humility.

Worship is that place where we remember that nonviolence aims for truth and not for power. The rhetorical cloak of nonviolence can be used to hide the will to power, which is the very foundation of violence. Nonviolence does not try to overcome adversaries by defeating them, but by convincing them. Nonviolence turns adversaries into friends, not by winning over them, but by winning them over. Nonviolence is a patient love that the Bible describes as "enduring all things."

In worship, we recall the clear words of God's judgment and mercy that comprise the prophetic vocation. Politicized theology is no substitute for prophetic witness. Prophets do not attack some injustices and ignore others; prophecy is profoundly anti-ideological. Prophets in the Bible are not known for smugness, pride, or bitterness. The biblical prophets loved the people and claimed them as their own; they spoke hard words with broken hearts. Their response to faithlessness was grief, not indignation. They knew that they too were sinners; only sinners make good prophets.

These idols still tempt us, but our worship can reveal and diminish their power over us.

Worship exposes all our attempts to make idols of ourselves. As the Bible says, "None is righteous, no, not one" (Rom. 3:10). Not even the most radical of Christians can save themselves. To trust God is to know that the world has already been saved by Jesus Christ. It is to know that we cannot save the world any more than we can save ourselves. All our work is done only in response to Christ's work among us; we love because we are loved, and we forgive because we have been forgiven. To confront the idols closer to home is finally to claim nothing for ourselves but much for Christ. This is conversion in the deepest sense; and one of its most dependable signs is the persistent presence of joy.

Joy, peace, love—these are some of the fruits of the Spirit by which the people of God are known. They are the marks of the fullness of conversion and, in turn, they are the signs of the integrity of our worship. The power of worship releases us from the captivity of all idols, and the gifts of the Spirit melt our hearts of stone into hearts of flesh. This is the promise of old.

For I will take you from the nations, and gather you from all the countries, and bring you into your own land. I will sprinkle clean water upon you, and you shall be clean from all your uncleanness, and from all your idols I will cleanse you. A new heart I will give you, and a new spirit I will put within you; and I will take out of your flesh the heart of stone and give you a heart of flesh. And I will put my spirit within you, and cause you to walk in my statutes and be careful to observe my ordinances. You shall dwell in the land which I gave to your fathers; and you shall be my people, and I will be your God. (Ezek. 36:24–28)

Verse 27 here could be read: "I will fill you with the Holy Spirit and guide you in your worship." The Spirit and worship are two important and interconnected parts of the larger promise to God's people. In the New Testament, the "upbuilding" of community is almost always spoken of in connection with worship. A strong case can be made for the community gathered for worship being the context where the Spirit is most visible in building up the body of Christ. In the section of his *Church Dogmatics* on "The Holy Spirit and the Upbuilding of the Christian Community," Karl Barth writes: "It is not only in worship that the community is edified. . . . But it is here first that this continually takes place. If it does not edify itself here, it certainly will not do so in daily life, nor in the execution of its ministry in the world."[1] All of us have much to learn in recovering worship as the center and presupposition of the whole Christian life, the very atmosphere in which we live. Worship and the daily life of obedience of the Christian are not two separate spheres, but two concentric circles; worship is the inner circle that gives the outer circle of daily obedience its content and character.

Thus we may judge the value of worship, the inner circle, by looking at the shape of the outer circle, or the daily obedience it produces. Our worship should spread from the inner circle to the wider circle of our everyday lives as Christians; our daily speech, acts, and attitudes are ordained to be a wider and transformed worship. Likewise, we can judge the quality of our involvements in the world by looking at the inner shape of the worship experienced in our communities. The nature of our corporate worship will ultimately be the test of our other involvements in the world. The quality of our worship will reflect, if not determine, the quality of everything we do, including whether we will serve and minister rightly in the world. If we are not experiencing the power of God in our worship with each other, we will not experience the power of God in our involvements in the world.

The work of the Holy Spirit is central to ethical discernment. Throughout the book of Acts, the Spirit is active, especially in making decisions. If the proportionate space given to various themes in Acts is any indication, then the basic work of the Spirit is to guide in discernment; prophecy, testimony, inward conviction, and empowerment are less commonly mentioned works of the Spirit.[2] Oscar Cullmann argues that the precise function of the Spirit is best summed up in Paul's word *dokimazo,* translated variously as "testing," "discerning," "proving," or "determining": "The working of the Holy Spirit shows itself chiefly in 'discerning,' that is in the capacity of forming the correct Christian ethical judgment at each given moment. . . . This 'discerning' is the key to all New Testament ethics."[3] This argument may seem to conflict with our understanding that the key to all New Testament ethics is the kingdom of God and the teachings of Jesus. In the final analysis, though, there is no conflict. Biblical scholarship recognizes an inner connection in Paul's use of the words "Spirit" and "kingdom of God"; the role of the Spirit in Paul's teaching is similar to that of the kingdom in the Synoptic Gospels.[4]

Our communities must be climates of mutual trust and respect that will enable us all to hold in genuine unity the entire range of the Spirit's gifts. Jesus promised his disciples that he would send a "Comforter" to them. The Spirit would provide the knowledge and experience of God's life among them. The presence of the Spirit would supply the daily reality of the love they had been drawn into through Christ. The Spirit would seal their fellowship with God and Jesus Christ. Through the work of the Holy Spirit, the Bible promises,

God will raise up a faithful people to do what needs to be done. The origin of the church was at Pentecost when God poured out the Holy Spirit to fill the church and to empower its life and mission. The church is not expected to fulfill its mission in the world through its own strength or with mere human resources.

A biblical promise is that the people of God will be given the gifts necessary for their mission as they offer themselves to God in faithful obedience. Because the nature of our warfare with the principalities and powers of the world is at root a spiritual battle, the Christian community must learn not to fight them with only the same economic and political weapons that the powers have at their disposal. Rather, we must learn to employ spiritual resources and power and to use the gifts of the Spirit to resist the destructive designs of the principalities. The discovery of how to release and channel that kind of spiritual power in confrontation with institutional forces of injustice and violence is a central community undertaking; it is an undertaking that presumes a deep, historical rootedness in worship. Since there is no biblical restriction on manifestations of the Spirit, we may find new gifts awakening among us as we seek to be the church in the midst of social and political realities that Paul could not envision in the first century.

The Holy Spirit, therefore, is central to worship and to the building of Christian community, to our active involvement in the world and to our ethical discernment. The internal dynamic of the Spirit's activity anchors our worship and creates community. Many of us have come to see the building of Christian community as central to our lives and to our mission in the world. However, the experience of our own community, during its early days in Chicago, is probably far too typical.[5] We watched helplessly, with bewilderment and disillusionment, as all our highest dreams and noblest efforts to build community crumbled around us. There were many reasons for this: our lack of wisdom in handling interpersonal friction, a fear of leadership, a pride that kept us from learning from others. As we look back, though, perhaps the biggest reason was that we simply did not understand the centrality of the Spirit to building community. We had to open ourselves to an unclogging of the Spirit among us. And we had to learn how to praise the Lord.

THE WORDS "PRAISE THE LORD!" conjure up a wide range of feelings in different people. Some of us feel comfortable and at ease with acclamations of praise; others feel awkward and not a little embarrassed.

The language of praise brings with it jumbled feelings of fear and discomfort. We are afraid of the false piety; we sense that somehow these verbal explosions of praise gloss over the suffering of real people in a real world. We are wary, because we know of movements and groups that praise the name of the Lord while tending to ignore his teaching and commandments. We picture scenes of wealthy suburbanites enthusiastically praising God for their prosperity; or we imagine prayer groups in the Pentagon, assuring themselves of God's blessing during their lunch hour, then going back to planning for war.

But in our reaction to the piety that accepts injustice and violence, we are in danger of losing something vital. We risk losing something that the world needs perhaps more than ever. There is power in the simple praise of God. We must proclaim not only the kingdom but also Jesus, the author of the new order. The name of the Lord, and not just his justice and peace, is worthy of worship and honor. Jesus's followers should love the poor; they also should sing the praises of the God who loves the poor.

If we are immersed in the world's suffering and are feeling its pain, praise may be the only thing that will prevent us from being overcome by the world. As we daily share in the struggles for peace and justice, praise will become more, not less, necessary. The despair born of awareness will exhaust us unless we receive the freedom to praise God in the midst of it all. Praise is a language of freedom. It lifts us out of ourselves. In worshiping the Lord we focus on God and are liberated from our self-preoccupation. To praise God, we must lose our pride and gain a willingness to seem foolish and childlike; we must let go of our self-images and trust the Spirit.

Praise is a language of trust and identity. It affirms the trustworthiness and dependability of God and asserts our reliance on God. Praise also requires that we trust our brothers and sisters. As long as we fear others will judge us or think us silly for offering free expressions of worship, praise will die in our throats. Praise establishes who is who. The Lord is God; God made us, and we are God's people. To praise the Lord is to acknowledge God as our shepherd and ourselves as sheep. That can be a blow to our self-images. Worship and praise affirms that we need the shepherd's care, that we are completely dependent on the Lord.

Praise is the language of politics. Our political identity is rooted in the praise of God, not in any ideology or system. We are freed to act in the world in a new and revolutionary way. No longer defined or constricted by ideology, our witness can truly challenge the world's assumptions and resist co-optation by any party or institution. The authority of every other god melts away in worship of the Lord. Praise reserves the vocabulary of absolute value and ultimate significance for God alone. It puts the world's politics in proper perspective. We are allowed to enter political struggles free of false expectations, loyalties, and illusions. Our public witness is for the praise of God and not for the glory of human institutions and ambitions.

A church of praise is a church that places itself under God's care and sovereignty. The church acknowledges its highest loyalty and celebrates its deepest love. Such praise is at the heart of our opposition to tyranny and violence. As Jesus and his followers in the early church discovered, such praise often moves the gods of the state to jealousy, anger, and retribution. Praising the Lord declares a freedom that threatens the status quo. The praise of God joyfully announces that we no longer see the world as the world would have us see it; we see God in and through all of creation.

The church is to be, of necessity, a community of resistance to evil and injustice. But the resistance flows from a sense of joyful praise and loving celebration. We are first called to be a community of celebration, celebrating the life to which we are called, the life given to us to share together and then give away for the sake of the world. We become a community of resistance. But before that, through that, and even after that, we are a community of praise and celebration.

Worship is to resistance as a tree's roots are to its branches. Among the dictionary definitions of *roots* are the following: "a means of anchorage and support," "an underlying support or basis," "the essential core or heart." Conversion always involves returning to our roots, to our first love. Thus all tests of conversion are tests of whether or not we remember who we are and to whom we belong. We have mentioned the biblical cycle of forgetfulness among the people of God, how it leads to falling away and needing to be called back, to be converted. To worship is to remember who we are in the midst of everything that would have us forget. Worship is nothing more than recalling our roots; it is nothing less than reclaiming our identity as God's people.

I recall the Passover scene from the Broadway musical and film *Fiddler on the Roof*: The father and mother gather all the children. They are all dressed in their best clothes. The table is set properly and everything is just as it is supposed to be. The faces of the children ask the question, "Why are we doing all this?" Their eyes, especially those of the younger ones, say, "I don't understand. Why do we go to all this trouble?" The answer, in the parents' eyes and in everything they are doing, is, "Because we are Jews." They are Jews in Russia. They are in danger, but they will not forget the fact that they are Jews. They do all this to remember who they are. They do it because their fathers and mothers did it before them, beginning at a time when they, too, were too young to understand fully. The scene stretches back into past generations, and it will reach forward into future generations. These children will someday answer the inquisitive eyes of their children, saying, "We do this because we are Jews."

They understood the importance of maintaining their identity in the world. That is the role of worship. No matter what else happens, we can never forget that we are God's people. We must do the things that remind us, that recall our history. We remember our tradition and God's action in the world on our behalf. We tell the story again and again because, in our remembering, everything is at stake.

The early Christians also understood that their very survival in the world depended on the clarity of their identity with God and with God's purposes for them in history. Because their identity was secure in their Lord, they were a people who could say, "You may do as you will, but as for us, we will live as God's people."

This need and desire to live as God's people greatly affected the evolution of worship in Sojourners community. The celebration of the Lord's Supper gradually moved to the center of our corporate life, despite the fact that many of us came from church backgrounds where communion was not often observed. Regardless of religious training, the experience of Christian community seems to create fellowships where the celebration of the eucharist is central. The eucharist graphically reminds us of what Christ has done for us. Communion roots us in the love of Christ. At the table, we are reminded of that love, which is the very basis of our life together. The eucharist is the drama of remembering who we are as God's people. Jesus says, "Do this in remembrance of me." Why are we told to eat this bread and drink this wine over and over again? Because we are so forgetful. We need to be reminded again and again.

The origins of the eucharist are in the Passover meal. The keeping of the Passover reminded the Jewish people of their deliverance from the Egyptians and of God's continual faithfulness. In the midst of the Passover meal, Jesus instituted the breaking of the bread and the taking of the cup to remind the disciples of his suffering and death for them. Jesus did not want them to forget how he gave his life and, as his followers, what their vocation would be. Every time we partake of the bread and the wine, we recall Jesus's way and remember what our way is to be.

Have this mind among yourselves, which you have in Christ Jesus, who, though he was in the form of God, did not count equality with God a thing to be grasped, but emptied himself, taking the form of a servant, being born in the likeness of men. And being found in human form he humbled himself and became obedient unto death, even death on a cross. (Phil. 2:5–8)

This is the meaning of the incarnation. It is the heart of Christian faith. Jesus is God made vulnerable. God has come to share our flesh, our life, our human lot. God has drawn near to us so that we may understand and stand close by the one who loves us. Clarence Jordan, when he described the incarnation, liked to say, "God moved in with us."

The Christian answer to human hurt is not fixed doctrine; we do not solve problems with speculative theology; the Christian response to oppression is not moral philosophy. Our God does not render a detached pronouncement on our situation. Rather, God comes to join us, to enter into our circumstances, to feel what we feel, and to walk with us. The name Emmanuel means "God with us." Jesus gave up his divine prerogatives, becoming one of us in order to show us the way. The Philippians passage is known as the *kenosis,* the self-emptying of Jesus. God became a servant among men and women. The love that established the pattern of servanthood in Jesus Christ would forever be the heartbeat of Christian faith.

We have spoken of a world painfully divided between rich and poor, powerful and powerless. The word *liberation* expresses the deepest aspiration in the hearts of the poor and powerless around the world. The human spirit desires freedom. Once that energy is set loose, nothing will stop it.

The word that best describes the posture of the rich and powerful people of this world is *control.* Burdened with anxiety, the affluent cling even more desperately to their control of resources, privileges, and power. In the face of the momentum of the poor toward liberation, they seek further fortifications for their position of dominance. The energy of those at the top of the world order is increasingly invested in a grim determination to hang on to control at all costs, even at the cost of total war.

The relationship of the poor to the rich today can be illustrated by two hands. An outstretched hand is rising up from the South, seeking liberation. As the hand moves outward, it encounters another hand, made into the palm-forward sign to halt; it is a wall blocking the movement of the outstretched hand. The open hand moves against the wall, again and again, gradually becoming a fist pounding on the hand that refuses to yield. Soon the second hand also clenches and hardens into a fist. This is how we must characterize the world today, as the two fists pounding against each other.

The eucharist stands between the two fists. In it we see the cross. For the poor, the cross is the promise of liberation and the picture of what it will cost. The cross is empowerment; that is its hope for the downtrodden. But the eucharist also says that deliverance is not cheap. Suffering will pave the road to freedom. In the bread and in the wine, the poor can see the face of the one who has gone before, the one who has paid the price for their salvation, the one who knows their suffering and pain and will stand with them in it.

For the rich, the cross is also the promise of liberation and the demonstration of the only way it can occur. The cross is relinquishment; that is its hope for those in control. The eucharist reveals the one who gave up everything to accomplish our salvation. Through the one who has relinquished all, we can see the possibility and the power of our own relinquishment. Releasing control will allow us to lay down the heavy baggage that has become an obstacle blocking our way to freedom. Embracing the cross means following the example of Jesus: opening our hands and becoming humble servants. Choosing the way of the cross means losing control and gaining new life.

The relinquishment of the cross is the gift of Christ to those who are now in control of human relationships. Relationships of control and domination can become relationships of trust and cooperation: the rich in relation to the poor; the powerful in relation to the powerless; the strong in relation to the weak; those of the North in relation to those of the South; those who are white in relation to those who are black, brown, red, and yellow; men in relation to women; those with military superiority in relation to those who are the targets of their weapons.

Again, we can see this with our hands. Relinquishment means no longer to grasp, to hold back, to push away, to hold on with all of our might. Rather, it means to see Jesus on the cross with open hands; it means to see him now standing before us in the eucharist, his open hands reaching out to us, inviting us to open our hands. An open hand cannot be held by a fist; an open hand can be embraced only with another open hand. Our clenched fists will be slowly opened in the presence of the eucharist. Here, in the meal that reminds us of Jesus's vulnerability, closed fists can be opened and joined in the handshake of fellowship. Empowerment and relinquishment meet at the foot of the cross. The eucharist is the sign of hope for the controlled and the controlling. At the table where we celebrate the Lord's life broken and poured out for all, liberation is freely offered to both.

In the eucharist, we see the vulnerability of God. We are

taught the all-important lesson that our own vulnerability is the way to identify with Christ in the world. The regular celebration of this holy meal establishes and sanctions the posture of vulnerability for the church's life in the world.

Christian community is the arena in which we learn the way of the cross, experience vulnerability, and see the path for the church's future. Our life together and the offering of it for the world will teach us, often painfully, the meaning of Jesus's way. The eucharist stands in the center of our community life to picture all that for us, to explain what we are experiencing and why, to assure us that what we now experience is a part of God's plan, and to remind us that this is the way the world will be saved. The eucharist is for our sake, and we desperately need it. It becomes a place of anguish, where the pain of letting go is deeply felt; a place of great joy and thanksgiving as we recognize what God has done and is doing among us; a place that absorbs our hate and heals our fears; a place that sets us free to give our lives away boldly for the sake of the kingdom.

The eucharist reminds us that the power of God is released at the very point of our risk-taking, our letting go. It teaches us to look for God in the places where our faith is being stretched, where we are assuming risks. The eucharist invites us to extend to the world the vulnerability we are learning together in Christ. It makes possible our personal vulnerability and enables us to see how our political vulnerability is the prerequisite for peace and reconciliation in a world of conflict. Jesus is in the eucharist, calling us and inviting us to let go, to trust him, to follow him, to lose ourselves in him, and finally to be free in him.

The eucharist is the only ritual that Jesus asked us to repeat again and again. Jesus wants us to remember him—not in triumph, power, and might, but in being broken, poured out, and emptied. In the broken bread and the outpoured wine, he bids us to remember him in this particular way.

The fists we have clenched must become open hands. If we would embrace Jesus, if we would see in the eucharist the road to our freedom and wholeness, then we would know the power of conversion to open our hands and soften our hearts. We would know that he who emptied himself, he who became poor in things so that we might become rich in spirit, has indeed been lifted up.

> Therefore God has highly exalted him and bestowed on him the name which is above every name, that at the name of Jesus every knee should bow, in heaven and on earth and under the earth, and every tongue confess that Jesus Christ is Lord, to the glory of God the Father. (Phil. 2:9–11)

CHAPTER 7

The Victory

But the angel said to the women, "Do not be afraid; for I know that you seek Jesus who was crucified. He is not here; for he has risen, as he said. Come, see the place where he lay. Then go quickly and tell his disciples that he has risen from the dead, and behold, he is going before you to Galilee; there you will see him."

Matthew 28:5–7

JESUS IS ALIVE. That was the rumor that spread through Jerusalem that first Easter morning. Women came to the tomb early in the morning, the first witnesses to the resurrection. Their testimony as women was not even admissable in court under Jewish law; the word of a woman had no public credibility in that patriarchial culture. But God chose to reveal the miracle of Jesus's resurrection first to women. They were told to report the astonishing news of the empty tomb to the men. At first, the men did not believe it.

Jesus's first appearance was also to a woman, Mary Magdalene. She was in the garden near the tomb, stricken with grief. The one who had accepted and forgiven her, the one whom she loved so deeply, was gone. She saw a figure she thought was the gardener and said to him, "They have taken my Lord. Do you know where they have laid him?" Then a familiar voice called her name, "Mary." She looked up and recognized him. "Master!" she cried. Her Lord had come back, and the heart of the woman who had been cleansed by his love leapt for joy. Mary went straight to the disciples with a simple testimony, "I have seen the Lord." Their excitement must have been enormous.

The disciples were in hiding behind locked doors from fear of the authorities, says the Bible. They had seen what had happened to their leader and were afraid they would be next. So they huddled in secret.

The ones at the tomb who appeared as "young men in shining garments" told the women to go tell the disciples *and Peter*. Peter had always been the leader among the disciples, but he had betrayed his Lord three times with oaths and curses. Peter denied his Master from fear. The strong fisherman wept bitterly and became utterly dejected after the death of the Lord. Jesus especially wanted Peter to know of his resurrection. He wanted to make sure Peter was told, not as a rebuke, but so Peter would know that he was alive and that he still loved him. When the women told them the news, Peter and John ran to the tomb. John, younger and faster than Peter, arrived first and waited at the entrance, peering into the darkness. Peter, always the impulsive disciple, didn't stop at the entrance; he went right inside. He had to see. He had to know. They saw the empty tomb, and they believed.

Then there were the two disciples on the road to Emmaus. They didn't recognize Jesus until he broke the bread. They also rushed to tell the disciples. Imagine the situation. The air was electric with rumors and reports of witnesses who said they had seen him. Most of the disciples had not yet seen him and were full of wonder. Could it be? It was too good to be true. A world that had ended for them three days earlier now seemed to be opening again.

Then Jesus came and stood among them. "Peace be with you," he said, as he looked into their eyes. Think what they must have felt at that moment. He showed them his hands and his feet. "It is I, myself . . . touch me and see." They could hardly believe what they were seeing. He even took a fish and ate it, just to show them he was real. He recalled to them the Scriptures and his own foretelling of his death and resurrection. It was really he, and he was really alive.

Thomas wasn't there. When the others told him, he didn't believe it. Perhaps wounded with pain and disillusionment, perhaps filled with bitterness and cynicism, Thomas would not let his hopes be rekindled. He said, "Unless I see the marks of the nails in his hands, unless I put my fingers in the place the marks were, and my hand into his side, I will not believe."

Later, Jesus came to his disciples again. This time, Thomas was present. "Thomas," he said, "put your finger here and see my hands. Put out your hand and place it in my side. Do not be faithless, but believing." Thomas must have witnessed the marks of Jesus's suffering with tears in his eyes. "My Lord and my God," he humbly exclaimed. For Thomas, and for them all, unbelief was turned to belief when they saw their Lord and the marks of his suffering. They were converted by the resurrection.

The disciples had left everything to follow Jesus. He had touched their lives as no one else ever had. He was the one who loved them, and the one whom they had grown to love. Jesus was alive again and among his disciples as before, but now in a new way. The first words spoken to Jesus's followers at his empty tomb were, "Do not be afraid. . . . He is not here; for he has risen, as he said. Come, see the place where he lay." And the Scriptures say, "When they saw the Lord they were filled with great joy."

Jesus of Nazareth was delivered up by the chief priests and killed by the Romans under Pontius Pilate. He was dead and, three days later, was alive again. A man who died had been raised from the dead. History has been able to offer no other believable answer to the fact of his empty tomb.

The guards who had been posted at the tomb ran to tell the chief priests what had occurred. Their very lives were at stake for failing to prevent the tomb from being opened. To break the Roman seal that had been placed at the entrance to the tomb was against the emperor's law and punishable by death. The resurrection of Jesus Christ was, then, an act of civil disobedience. The chief priests agreed to protect the guards if they would go along with a story they made up, saying that the disciples had stolen the body.

But the story failed. Something had happened and the disciples had lost their fear. A dejected and defeated band was filled with faith and confidence. They had seen the Lord, and they had been converted.

When the disciples saw Jesus, they came out of hiding. Until then, they had been cowering behind closed doors, controlled by fear. They had feared the Jewish authorities and the Romans who stood behind them. They had feared the power of the soldiers, the courts, the temples. And they had been afraid of their own faithlessness and inadequacy.

Until they saw Jesus, the disciples viewed the world the way others did. The central reality of their lives had been the power of the system and their own powerlessness. But when they saw him, they unlocked the doors, came out, and began turning the world upside down. The disciples were converted; they knew another reality then, one that was truer, greater, stronger, and a more compelling authority than the realities that had paralyzed them with fear. Jesus had risen, and Jesus was Lord.

Today, Jesus stands among us, with the marks of his suffering plainly visible. He knows us, he knows our fears. We are afraid of economic hardship and diminishing resources; of the enmity between black, white, red, brown, and yellow peoples; of the volatile gulf between rich and poor; of the hurt between men and women; of violence stalking on every side; of the drift toward endless war; and of the ways that restoring broken fellowship might disrupt our lives and our security. We fear for ourselves and for our children. Like the disciples, we are afraid of the power of the systems of the world with their armies, their courts, their prisons, their threats. Like them, too, we fear our own powerlessness, weakness, and sense of inadequacy. We are insecure, frightened of our own emotions, and wary of trusting one another. We feel both the guilt of our sin and the vulnerability of our broken places. Above all, we fear pain, suffering, and finally death.

We, too, are hiding behind locked doors and are afraid to come out. Jesus knows our fears. He wants us to know his resurrection. He says, "Go, tell my disciples that I have risen and that I am going before them. And go tell . . ."—he slowly repeats each of our names. Tell him, tell her that we need not be afraid anymore. Like Peter, we have betrayed Christ because of our fears. But Jesus didn't hold Peter's fear against him. Nor does he hold our fears against us. We, too, have doubted like Thomas. We have become cynical, skeptical, and faithless. But Jesus stands among us, shows us his hands and his side, and he tells us to reach out and touch him. He tells Thomas and he tells us not to doubt but to believe.

Jesus died for our sins, our doubts, and our fears. He rose from the grave to demonstrate his victory over them and to set us free from their power. He wants us, like Peter, Thomas, Mary, and the others, to know his resurrection. He wanted them to know, and he wants us to know, that his love for his disciples has no bounds, that he died to set us free, and that he rose from the dead to show us his way was true. "Be of good cheer, I have overcome the world."

A conversion was wrought in the disciples. No longer afraid, they fearlessly proclaimed his resurrection in the streets of Jerusalem. What had brought about this miraculous transformation? They had experienced the resurrected Christ, and the experience converted them. They had seen the Lord, and they believed. They turned from fear and turned toward the Lord. Their lives became evidence of the resurrection.

Most people considered them fools for believing. The Jews did, the Romans did, the whole world did. But no one doubted that the disciples believed. The authorities told them to stop preaching, but to no avail. Peter, who had denied his Lord, now rose in the presence of the Jerusalem multitudes to preach the good news of Jesus Christ.

What is the good news? When all that sin had done, or could ever do, was laid on Jesus, it did not overcome him. Death could not swallow him. The grave was denied its victory. The witness of history and of his followers is that "he is risen." He is alive. He has triumphed over all. He is the victor over every sin, hate, fear, violence, and death. Nothing is stronger than his victory—nothing past, nothing present, and nothing future.

The Crucified One has prevailed over every principality, power, and dominion. He has "disarmed" them, made a "public example" of them, "discarded them like a garment," and "led them captive" in his victory procession (Col. 2:15). He has unmasked their illusions, exposed their lies, and showed them for what they are. He stands free of their threats, power, and control. He defeated them by letting them do their worst to him, then he vanquished them by the power of God's love and truth—weapons stronger than all the weapons of the world. Jesus bore the full weight of the world's sin, and he overcame it all. Today, as then, his people come together to confess, "Christ has died, Christ is risen, Christ will come again."

The resurrection always exists in relationship to the cross. The cross of Jesus, which appeared to be a complete defeat and utter failure, is revealed as the very means of the kingdom. The cross is, for us, not only the symbol of atonement for our sins but the pattern of our lives. Jesus moved toward it and provoked it when he could have chosen otherwise. "Bearing our cross" is more than simply enduring difficult personal circumstances. For Jesus, the cross was the expected result of a moral clash with the powers of his society. His cross, therefore, not only frees us from personal sin; it also liberates us from the power of this world. Living freely in relationship to those powers, establishing a moral independence from them, will ultimately lead to a cross. The cross is the sign of that freedom. The resurrection seals the truth of the cross; it declares that, once and for all, oppression and death are swallowed up in Christ's victory.

If this is God's way of salvation and liberation, do we have the right to choose any other way for relating to the world? The greatest offense in this world is the love that was willing to go to a cross in order to save the world. That was the political stance of God's suffering servant, a stance that was vindicated by the resurrection of Jesus Christ.

What about you and me today? Do we still doubt that this kind of love makes much sense in our complex technological world? Does the way of suffering servanthood seem out of place in our world of huge and powerful institutions? That doubt was the experience of the disciples between Good Friday and Easter Sunday. They, too, felt overwhelmed by the powers and forces that ruled the day. But they were converted. The disciples became the people of the resurrection. They began to live lives filled with the fruits of conversion. They began living in the power of the resurrection. We, too, can know the power of Christ's resurrection.

But such power will not come simply by attesting to the theological fact of the resurrection. We, like the first disciples, must come out of hiding and see the risen Lord. Seeing is believing, and believing is knowing that we must turn and follow Jesus. The wisdom of God in Jesus Christ will then be made known to the principalities and the powers through the church. The place in which the dominion of the powers is broken is the fellowship of the resurrection, the church community that nurtures growing confidence in the power of God's love. The resurrection restores broken fellowship. Because the Lord is risen, love reigns where fear once controlled.

To doubt or be suspicious of the power of that love is to doubt the resurrection. The doubting, suspecting parts of our lives are yet unhealed because we have not been converted; we have still to receive the love that would heal us and change the world.

We are invited to celebrate the victory of Christ and to live in the world showing we believe it to be true. We are invited to experiment with its truth by risking our security, comfort, resources, time, energy, and our very lives for the sake of his victory. We are called to be those who have left all and risked everything in testifying to his victory. We are called to demonstrate to the world, with our lives, that we have been converted by the resurrection.

The basis of our faith is that Jesus Christ is stronger than any of the powers that confront us: political tyranny, economic oppression, the logic of war. We confront the world's powers not merely with our own strength, resources, ideas, commitment, work, or resistance. Rather, we confront the world with the very life of the resurrected Christ among us.

Like Jesus's other disciples, who were controlled by their fear, we can be converted by seeing the resurrected Christ. Whenever we act in obedience to Christ, we are demonstrating his victory. Every time we act upon Jesus's lordship in our lives, are reconciled to a brother or sister, refuse to be controlled by the economic system, deny the absolute authority of the state, claim Christ's freedom over our fear, tear down the walls of race, class, and sex, love our enemies, stand with the poor, or resist the violence of the nations by acting for peace—we are demonstrating the victory of Christ in the world. His victory is present wherever it is claimed and acted upon. We are those who evaluate themselves and the whole world in the light of Christ's victory.

Preaching the cross and the resurrection of Jesus is foolishness to those who perish. Yet, fools for Christ formed the early church. As that tiny band of believers grew, the world could see the power in such foolishness.

That same foolishness is the only hope we have of breaking free from the present realities that so gravely threaten us. Only in the recognition of something that is more real can we see their authority as unreal. The greatest threat to any system is the existence of fools who do not believe in the ultimate reality of that system. Indeed, the first step in making new realities possible is to break free from the grip and the authority of the old realities. To repent and to believe in a new reality— that is the essence of conversion. We join the body of Christ whose purpose is to make visible this new reality in the world.

Without the resurrection, the defeated followers of Jesus would have simply faded away. He would have been just another prophet who was killed. But the resurrection vindicated the cross and validated the way of Jesus, establishing the authority of his Lordship. At the same time, the resurrection invalidated the authority of the system. It showed the world's way to be a lie. The world's definition of reality crucified Jesus. His resurrection proved that definition of reality to be false. Our system, too, has its definitions of reality—national security, economic expansion, political realism. The way of Jesus is thought to be as foolish today as it was in his day. His kingdom is totally alien to the present world order.

If we believe the resurrection, the world will consider us unreasonable, unrealistic, irresponsible, and irrelevant. A world full of incredible foolishness, of myths called "truths" and of lies called "logic," will insist that we are the fools. Yet the resurrection convinces us of God's wisdom. We finally know God's power to heal us and to transform our relationship to the world.

On Easter morning, and each day of our lives, we celebrate the reality of the resurrection of Jesus Christ, which triumphs over every other reality. In the face of the world and its systems, we proclaim the resurrection, saying, "We have seen the Lord." We see him in the lives of our brothers and sisters. We discover him in the faces of the poor, in the faces of all the victims, and in the faces of our children. We see him in the lives of Christians who have suffered and died because they believed. And we see the Lord in the bread and the wine. He shows us, as he did his disciples, the evidence of his suffering. He invites us to reach out, take, eat, and drink; he wants us to remember him, to see him, and to know his victory.

His way is life. The world's way is death. We can now stand

before the world's false realities and securities, free to deny them, denounce them, and remove ourselves from them. We stand before the reality of the resurrection and confess with the first disciples that Jesus is the Christ, the Son of God.

We stand before the world as fools. We are foolish enough to believe that Jesus's way is stronger and truer than the way of the world. We rest secure in the knowledge that he has, and will, overcome. We are called to be fools for Christ, a people saved by his cross and converted, finally, by his resurrection.

May God convert us to such foolishness.

Bibliography

CHAPTER I

Bonhoeffer, Dietrich. *The Cost of Discipleship*. Revised edition. New York: Macmillan Co., 1963.

Campbell, Will D. *Brother to a Dragonfly*. New York: Continuum, 1977.

Cassidy, Richard J. *Jesus, Politics and Society: A Study of Luke's Gospel*. Maryknoll, N.Y.: Orbis Books, 1980.

Conn, Walter E., ed. *Conversion: Perspectives on Personal and Social Transformation*. Staten Island, N.Y.: Alba House, 1978.

Costas, Orlando E. "Conversion as a Complex Experience: A Personal Case Study." In *Down to Earth: Studies in Christianity and Culture*, edited by John R. W. Stott and Robert Coote. Grand Rapids, Mich.: Eerdmans, 1980.

Day, Dorothy. *The Long Loneliness*. Reprint. San Francisco: Harper & Row, 1981.

_____. *Loaves and Fishes*. New York: Curtis Books, 1963.

Dayton, Donald W. *Discovering an Evangelical Heritage*. New York: Harper & Row, 1976.

Ellul, Jacques. *The Presence of the Kingdom*. New York: Seabury Press, 1967.

Fackre, Gabriel. *Word in Deed: Theological Themes in Evangelism*. Grand Rapids, Mich.: Eerdmans, 1975.

Green, Michael. *Evangelism in the Early Church*. Grand Rapids, Mich.: Eerdmans, 1970.

Jordan, Clarence. *Sermon on the Mount*. Revised edition. Valley Forge, Va.: Judson Press, 1952.

_____. *The Substance of Faith and Other Cotton Patch Sermons*. New York: Association Press, 1972.

Kraybill, Donald B. *The Upside-Down Kingdom*. Scottdale, Pa.: Herald Press, 1978.

McArthur, Harvey. *Sermon on the Mount*. New York: Harper & Brothers, 1960. Especially see "The Sermon and Ethics."

Newbigin, Lesslie. *The Finality of Christ*. Richmond, Va.: John Knox Press, 1969.

Shank, David A. "Towards an Understanding of Christian Conversion," *Mission-Focus* (Mennonite Board of Missions, Elkhart, Ind.), November 1976.

Sojourners. "Dorothy Day and the Catholic Worker." Vol. 5, no. 10 (December 1976).

_____. "The Legacy of Clarence Jordan." Vol. 8, no. 12 (December 1979).

Wallis, Jim. *Agenda for Biblical People*. New York: Harper & Row, 1976.

Yoder, John Howard. *The Politics of Jesus*. Grand Rapids, Mich.: Eerdmans, 1972.

CHAPTER 2

Berkhof, Hendrik. *Christ and the Powers*. Scottdale, Pa.: Herald Press, 1962.

Ellul, Jacques. *False Presence of the Kingdom*. New York: Seabury Press, 1972.

_____. *The New Demons*. New York: Seabury Press, 1973.

_____. *Propaganda: The Formation of Men's Attitudes*. New York: Vintage Books, 1965.

Gilkey, Langdon. *How the Church Can Minister to the World Without Losing Itself*. New York: Harper & Row, 1964.

Linder, Robert D., and Richard V. Pierard. *Twilight of the Saints: Biblical Christianity and Civil Religion in America.* Downers Grove, Ill.: InterVarsity Press, 1978.

Moberg, David O. *The Great Reversal: Evangelism and Social Concern.* Revised edition. New York: J. B. Lippincott Co., 1972.

Pierard, Richard V. *The Unequal Yoke: Evangelical Christianity and Political Conservatism.* Philadelphia: Lippincott, 1970.

Stringfellow, William. *Conscience and Obedience: The Politics of Romans 13 and Revelation 13 in Light of the Second Coming.* Waco, Tex.: Word Books, 1977.

_____. *An Ethic for Christians and Other Aliens in a Strange Land.* Waco, Tex.: Word Books, 1973.

_____. *A Public and Private Faith.* Grand Rapids, Mich.: Eerdmans, 1962.

CHAPTER 3

Barnet, Richard. *Economy of Death.* New York: Atheneum Publishers, 1969.

_____. *The Lean Years: Politics in the Age of Scarcity.* New York: Simon & Schuster, 1980.

Barnet, Richard, and Ronald Muller. *Global Reach: The Power of the Multinational Corporations.* New York: Simon & Schuster, 1974.

Birch, Bruce C., and Larry L. Rasmussen. *The Predicament of the Prosperous.* Philadelphia: Westminster Press, 1978.

Boff, Leonard. *The Way of the Cross—Way of Justice.* Maryknoll, N.Y.: Orbis Books, 1980.

Cone, James H. *God of the Oppressed.* New York: Seabury Press, 1975.

Daly, Herman E. *Steady-State Economics.* San Francisco: W. H. Freeman & Co., 1977.

Domhoff, G. William. *The Powers That Be: Processes of Ruling Class Domination in America.* New York: Vintage Books, 1978.

Ellul, Jacques. *The Technological Society.* New York: Vintage Books, 1954.

Goulet, Denis. *A New Moral Order: Development Ethics and Liberation Theology.* Maryknoll, N.Y.: Orbis Books, 1974.

Gutierrez, Gustavo. *A Theology of Liberation.* Maryknoll, N.Y.: Orbis Books, 1973.

Kirk, David. *Quotations from Chairman Jesus.* Springfield, Ill.: Templegate Publishers, 1969.

Longacre, Doris Janzen. *Living More with Less.* Scottdale, Pa.: Herald Press, 1980.

Malcolm X and Alex Haley. *The Autobiography of Malcolm X.* New York: Grove Press, 1964.

McGinnis, James B. *Bread and Justice: Toward a New International Economic Order.* New York: Paulist Press, 1979.

Miguez Bonino, Jose. *Christians and Marxists: The Mutual Challenge to Revolution.* Grand Rapids, Mich.: Eerdmans, 1976.

_____. *Doing Theology in a Revolutionary Situation.* Philadelphia: Fortress Press, 1975.

Mills, C. Wright. *The Power Elite.* New York: Oxford University Press, 1956.

Miranda, Jose. *Marx and the Bible: A Critique of the Philosophy of Oppression.* Maryknoll, N.Y.: Orbis Books, 1974.

Nelson, Jack A. *Hunger for Justice: The Politics of Food and Faith.* Maryknoll, N.Y.: Orbis Books, 1980.

Rifkin, Jeremy, with Ted Howard. *The Emerging Order: God in an Age of Scarcity.* New York: G. P. Putnam's Sons, 1979.

_____. *Entropy: A New World View.* New York: The Viking Press, 1980.

Schumacher, E. F. *A Guide for the Perplexed*. New York: Harper & Row, 1977.

_____. *Small Is Beautiful: Economics as if People Mattered*. New York: Harper & Row, 1973.

Sider, Ronald J. *Rich Christians in an Age of Hunger: A Biblical Study*. Downers Grove, Ill.: InterVarsity Press, 1977.

Sloan, Robert B. Jr. *The Favorable Year of the Lord: A Study of Jubilary Theology in the Gospel of Luke*. Austin, Tex.: Schola Press, 1977.

Walsh, William J., and John P. Langan. "Patristic Social Consciousness—The Church and the Poor." In *The Faith That Does Justice: Examining the Christian Sources for Social Change*, edited by John C. Haughey. New York: Paulist Press, 1977.

Williams, William Appleman. *Empire as a Way of Life*. New York: Oxford University Press, 1980.

Wilmore, Gayrand S., and James H. Cone, eds. *Black Theology: A Documentary History, 1966–1979*. Maryknoll, N.Y.: Orbis Books, 1979.

CHAPTER 4

Aldridge, Robert C. *The Counterforce Syndrome: A Guide to U.S. Nuclear Weapons and Strategic Doctrine*. Washington, D.C.: Institute for Policy Studies, 1978.

Aukerman, Dale. *Darkening Valley: A Biblical Perspective on Nuclear War*. New York: Seabury Press, 1981.

Bainton, Roland H. *Christian Attitudes Toward War and Peace*. Nashville: Abingdon, 1960.

Barnet, Richard J. *The Giants: Russia and America*. New York: Simon & Schuster, 1977.

_____. *Real Security: Restoring American Power in a Dangerous Decade*. New York: Simon & Schuster, 1981.

_____. *Roots of War: The Men and Institutions Behind U.S. Foreign Policy.* New York: Penguin Books, 1971.

Berrigan, Daniel. *The Discipline of the Mountain: Dante's Purgatorio in a Nuclear World.* New York: Seabury Press, 1979.

Cooney, Robert, and Helen Michalowski, eds. *The Power of the People: Active Nonviolence in the United States.* Culver City, Calif.: Peace Press, 1977.

Douglass, James W. *Lightening East to West.* Portland, Ore.: Sunburst Press, 1980.

_____. *The Non-Violent Cross: A Theology of Revolution and Peace.* New York: Macmillan Co., 1966.

Durland, William, ed. *People Pay for Peace: A Military Tax Refusal Guide.* 2nd revised edition. Colorado Springs: The Center on Law and Pacifism, 1980.

Durnbaugh, Donald F., ed. *On Earth Peace: Discussion on War/Peace Issues Between Friends, Mennonites, Brethren and European Churches, 1935–1975.* Elgin, Ill.: The Brethren Press, 1978.

Ellul, Jacques. *Violence: Reflections from a Christian Perspective.* New York: Seabury Press, 1969.

Hershberger, Guy Franklin. *War, Peace and Nonresistance.* Scottdale, Pa.: Herald Press, 1944; 3rd ed., 1969.

Hornus, Jean Michel. *It Is Not Lawful for Me to Fight: Early Christian Attitudes toward War, Violence and the State.* Revised edition. Translated by A. Kreider and O. Coburn. Scottdale, Pa.: Herald Press, 1980.

Kaplan, Fred. *Dubious Specter: A Second Look at the "Soviet Threat."* Washington, D.C.: Transnational Institute, 1977.

Kaufman, Donald D. *The Tax Dilemma: Praying for Peace, Paying for War.* Scottdale, Pa.: Herald Press, 1978.

Lasserre, Jean. *War and the Gospel.* Scottdale, Pa.: Herald Press, 1962.

Lens, Sidney. *The Day Before Doomsday: An Anatomy of the Nuclear Arms Race.* Boston: Beacon Press, 1977.

Lynd, Staughton, ed. *Nonviolence in America: A Documentary History.* New York: The Bobbs-Merrill Company, 1966.

Macgregor, G. H. C. *The New Testament Basis of Pacifism and the Relevance of an Impossible Ideal.* Nyack, N.Y.: Fellowship Publications, 1954.

Merton, Thomas. *Faith and Violence: Christian Teaching and Christian Practice.* Notre Dame, Ind.: University of Notre Dame Press, 1968.

Shannon, Thomas A., ed. *War or Peace: The Search for New Answers.* Maryknoll, N.Y.: Orbis Books, 1980.

Sharp, Gene. *The Politics of Nonviolent Action.* Boston: Porter Sargent Publishers, 1973.

Sider, Ronald J. *Christ and Violence.* Scottdale, Pa.: Herald Press, 1979.

Wolfe, Alan. *The Rise and Fall of the Soviet Threat: Domestic Sources of the Cold War Consensus.* Washington, D.C., Institute for Policy Studies, 1979.

Yoder, John Howard. *The Christian Witness to the State.* Newton, Kans.: Faith and Life Press, 1964.

_____. *Nevertheless: Varieties of Religious Pacifism.* Scottdale, Pa.: Herald Press, 1971.

_____. *The Original Revolution: Essays on Christian Pacifism.* Scottdale, Pa.: Herald Press, 1971.

CHAPTER 5

Bonhoeffer, Dietrich. *Life Together.* New York: Harper & Row, 1954.

Cosby, Gordon. *Handbook for Mission Groups.* Washington, D.C.: Potter's House Press, 1975.

Gish, Arthur G. *Living in Christian Community*. Scottdale, Pa.: Herald Press, 1979.

Jackson, Dave and Neta. *Living Together in a World Falling Apart: A Handbook on Christian Community*. Chicago: Creation House, 1974.

Jackson, Dave. *Coming Together: All Those Communities and What They're Up To*. Minneapolis: Bethany Fellowship Press, 1978.

Jones, James W. *The Spirit and the World*. New York: Hawthorne Books, 1975.

Pulkingham, Graham, et al. *Renewal: An Emerging Pattern*. Dorset, Eng.: Celebration Publishing, 1980.

Vanier, Jean. *Be Not Afraid*. New York: Paulist Press, 1975.

_____. *Community and Growth*. New York: Paulist Press, 1979.

CHAPTER 6

Berrigan, Daniel. *Uncommon Prayer: A Book of Psalms*. New York: Seabury Press, 1978.

Bloom, Anthony. *Beginning to Pray*. Maryknoll, N.Y.: Orbis Books, 1972.

Carretto, Carlo. *In Search of the Beyond*. Garden City, N.Y.: Image Books, 1975.

_____. *Letters from the Desert*. Maryknoll, N.Y.: Orbis Books, 1972.

Crosby, Michael. *Thy Will Be Done: Praying the Our Father as Subversive Activity*. Maryknoll, N.Y.: Orbis Books, 1977.

Doherty, Catherine de Hueck. *Poustinia: Christian Spirituality of the East for Western Man*. Notre Dame, Ind.: Ave Maria Press, 1975.

Ellul, Jacques. *Prayer and Modern Man*. New York: Seabury Press, 1970.

Foster, Richard J. *Celebration of Discipline: The Path to Spiritual Growth*. San Francisco: Harper & Row, 1978.

Merton, Thomas. *Contemplation in a World of Action*. Garden City, N.Y.: Image Books, 1973.

_____. *New Seeds of Contemplation*. New York: New Directions, 1961.

_____. *Contemplative Prayer*. Garden City, N.Y.: Image Books, 1969.

Nouwen, Henri. *Clowning in Rome: Reflections on Solitude, Celibacy, Prayer, and Contemplation*. Garden City, N.Y.: Image Books, 1979.

_____. *The Genesee Diary*. Garden City, N.Y.: Image Books, 1976.

_____. *Reaching Out: The Three Movements of the Spiritual Life*. Garden City, N.Y.: Doubleday, 1975.

Acknowledgments

When *Call to Conversion* was originally published, I thanked those who helped to shape the book—all named below in the original Acknowledgments.

For the revised edition, I would like to thank the following people. First, Mark Tauber, my publisher at Harper San Francisco, who pushed us to re-release *Call to Conversion,* believing its timeless message of discipleship was ready to be heard by a new generation. Second, my trusted and very competent editor at Harper San Franciso, Eric Brandt, made sure (as he always does) that everything was completed when it had to be done, while at the same time showing a remarkable sensitivity for all the other demands on the schedule. Grace and efficiency are not always gifts that live well together, but they do in Eric. And finally, this book (and all my books) would simply not get done without the research, editing, wise judgment, and tireless hours of work put in by my friend and associate, Duane Shank. In all my writing projects, Duane helps keep me going in the right direction and get there on time.

ACKNOWLEDGMENTS FOR 1981 EDITION

Thanks for this book must go to the whole Sojourners community. It literally grew out of our life together and was, indeed, a community endeavor. The experience of Sojourners shapes the book throughout, and it was the support and involvement of many that brought the book to completion.

A few people deserve special thanks:

Bob Sabath was the one who believed in the book from the start, who continually encouraged me to write it, and who was most involved in its initial conception and planning. His skillful help in outlining the material, researching, talking through basic questions, and working with the bibliography and footnotes proved invaluable.

Jim Stentzel came up with the good idea of my getting away for a few weeks to finish the book, and then offered to come along to help. The only thing more valuable than his help in editing, rewriting, and critical feedback was the spiritual companionship we shared during those weeks. The time was a unique combination of editorial collaboration, friendship, and worship rooted in a daily discipline of eucharist.

Lindsay Dubs did the initial copy editing on the book and served as a sort of managing editor for the whole project. Her characteristic positive and hopeful spirit often provided a needed spark and lift during the busy production schedule.

Cathy Stentzel offered her steady personal support and good judgment, and made many sacrifices to see the book completed. Her friendship, warm encouragement, and competent assistance helped bring much-needed order to my days.

Bob Raines and Kirkridge made possible a month of solitude to begin the project, and my parents, Phyllis and James Wallis, provided the quiet of their home to complete it. My editor at Harper & Row, Roy M. Carlisle, gave very practical help and suggestions along the way.

This book is dedicated to Bob Sabath, Jackie Sabath, and Joe Roos, who have been faithful since the beginning, have loved with their whole hearts, and have taught me the meaning of the word *community*.

Notes

CHAPTER 1

1 See the article on "Conversion" by F. Laubach, J. Goetzmann, and U. Becker, in *The New International Dictionary of the New Testament,* ed. Colin Brown, 3 vols. (Grand Rapids, Mich.: Zondervan, 1975–78), 1:353–362 (hereafter cited as *NIDNT*); and the article on "Strepho" by G. Bertram in the *Theological Dictionary of the New Testament,* ed. Gerhard Friedrich, trans. and ed. Geoffrey W. Bromiley, 10 vols. (Grand Rapids, Mich.: Eerdmans, 1964–1976), 7:714–29 (hereafter cited as *TDNT*). Both of these New Testament dictionaries discuss the Hebrew roots of conversion.

The Hebrew verb *shub* is basically one of motion and preserves much of the real meaning of conversion as an actual turning around, a *reversal from* and a *turning toward* something. The concept of conversion, however, is not exhausted by this one word. Walter Eichrodt shows that there are at least twenty common phrases used to indicate conversion and a return to God, including: to seek God, to humble oneself before God, to soften one's heart, to seek the good and hate the evil, to break up the fallow ground, etc. *Shub,* though, sums up all those other descriptions in a single, pregnant word.

> The metaphor was an especially suitable one, for not only did it describe the required behavior as a real act—"to make a turn"—and so preserve the strong personal impact; it also included both the negative element of turning away from the direction taken hitherto and the positive element of turning toward, and so, when combined with prepositions, allowed the rich content of all the many other idioms to be reproduced tersely yet unmistakably (*Theology of the Old Testament,* Vol. 2, trans. J. A. Baker [London: SCM Press, 1967], pp. 465–466).

2 See Laubach, Goetzmann, and Becker, "Conversion," *NIDNT,* 1:353–362; Bertram, "Strepho," *TDNT,* 7:714–29; J. Behm and E. Wurthwein, "Metanoia," *TNDT,* 4:975–1008. The process of conversion is expressed in the New Testament by three primary word groups, which deal with its various aspects: *epistrephein, metamelomai,* and *metanoein.* Since *metamelomai* expresses the feeling of sorrow for

sin without necessarily encompassing a turn to God, the other two terms become the predominant word groups that carry the central meaning of conversion in the New Testament. *Strephein* is the root for ten basic terms in the New Testament that refer to conversion and, in different contexts, may mean turn, return, turn around, turn back, be converted, change, turn away from, or conversion. *Metanoia* means conversion or repentance—in verb form, to repent or be converted.

The New Testament is full of other symbols that describe conversion. The fact that these two primary word groups do not occur often in Paul or John does not mean that the idea of conversion is not present there, but only that in the time between the writing of the Gospels and the epistles a more specialized terminology had developed. Both Paul and John convey the idea of conversion through the imagery of faith. Paul speaks of conversion as "being in Christ," as the "dying and rising with Christ," as the "new creation," or as "putting on the new man." John represents the new life in Christ as "new birth," as passing from death to life and from darkness to light, as the victory of truth over falsehood and of love over hate.

3   Karl Barth describes conversion as a twofold call to "halt" and then to "proceed" out of our sleep. Once awakened, we realize we are going down the wrong road and need to have our feet set upon a new one. Our former movement is halted, and we are told to proceed in another direction. The two movements of halting and proceeding belong together, says Barth, and form the essential unity of conversion. Karl Barth, "The Awakening to Conversion," in *Church Dogmatics*, trans. G. W. Bromiley (Edinburgh: T. & T. Clark, 1958), IV/2:560–561.

4   What did Jesus mean when he said "repent"? Joachim Jeremias tries to answer this question by looking especially at the parables. See his *New Testament Theology* (New York: Scribner's, 1971), p. 153. Jeremias notes that Jesus demands repentance in its breadth and depth by presenting a whole series of new pictures. The pictures are always concrete and specific to the person's situation. Jesus "expects the publican to stop cheating (Luke 19:8), the rich man to turn away from his service of Mammon (Mark 10:17–31), the conceited man to turn away from pride (Matt. 6:1–18). If a man has dealt unjustly with another, he is to make good (Luke 19:8). From hence forward, life is to be ruled by obedience to the word of Jesus (Matt. 7:24–27), the

confession of him (Matt. 10:32f.), and by discipleship that comes before all other ties (verse 37).

5   Gabriel Fackre, *Word in Deed: Theological Themes in Evangelism* (Grand Rapids, Mich.: Eerdmans, 1975), pp. 84–94. Fackre's chapter on conversion is a well-organized summary of the meaning of conversion as it is understood in this book. Conversion is primarily a turning, which always involves a "turning from something" and a "turning toward something." Conversion is turning to Christ in repentance and faith. *Epistrephein* always includes the element of faith. Since the connotation of *metanoia* is less broad (it refers to repentance), faith is often expressly used in a complementary way with it. If there is a distinction to be made between the two New Testament words, then *metanoia* emphasizes somewhat more strongly the element of turning away from the old, and *epistrephein* emphasizes turning toward the new (see Fackre, *Word in Deed*, p. 85).

6   This is only a partial listing of the innumerable threads that are part of the whole cloth of conversion. A more exhaustive list might include: from fear to hope, from spiritual blindness to the light of Christ, from idolatry to true worship, from control to relinquishment, from despair to joy, from wealth to simplicity, from the Bomb to the Cross, from alienation to reconciliation, from domination to servanthood, from anxiety to prayer, from false security to trust, from selfishness to sacrifice, from superiority to equality, from chauvinism to mutuality, from consumption to conservation, from accumulating to giving away, from hate to love of enemies, from swords to plowshares, from mammon to the God of the poor, from violence to peace, from exploitation to justice, from hardness of heart to compassion, from oppression to liberation, from individualism to community, from America first to Jesus first.

7   Laubach, Goetzmann, and Becker, "Conversion," *NIDNT*, 1:355. "Conversion involves a change of lords. The one who until then has been under the lordship of Satan (cf. Eph. 2:1f.) comes under the lordship of God, and comes out of darkness into light (Acts 26:18; cf. Eph. 5:8)."

8   See Walter E. Conn, ed., *Conversion: Perspectives on Personal and Social Transformation* (Staten Island, N.Y.: Alba House, 1978). The overall thrust of this book is that conversion is a progressive, integrative process that has consequences in society—not merely in the spiritual life of the individual. Conversion is not a single event, but an ongoing process in which the totality of a person's life is transformed. Also see

Orlando E. Costas, "Conversion as a Complex Experience: A Personal Case Study," in *Down to Earth: Studies in Christianity and Culture*, ed. John R. W. Stott and Robert Coote (Grand Rapids, Mich.: Eerdmans, 1980). In addition to showing that conversion is a continuous process, Costas emphasizes the "challenge of conversion inside the church.... In order to call others to conversion, it must be converted itself." Conversion is most often used for the turning of unbelievers for the first time to God (Acts 3:19; 26:20), but sometimes it is linked to erring believers (James 5:19f.) who are brought back into a right relation with God (see G. Bertram, "Epistrepho," *TDNT*, 7:727). Repentance (*metanoia*) can likewise be used of believers, and is found in reference to the problem of apostasy inside the church (Rev. 2:5, 16, 21, 22; 3:3, 19).

9   See Lesslie Newbigin, *The Finality of Christ* (Richmond: John Knox Press, 1969), pp. 93–94. Newbigin describes the historical characteristic of the call to conversion as found in the prophets and in John:

> It is a call to concrete obedience here and now in the context of the actual issues of the day.... Conversion is [not] some sort of purely inward and spiritual experience which is later followed by a distinct and different decision to act in certain ways. The idea that one is first converted, and then looks round to see what one should do as a consequence, finds no basis in Scripture. And yet this idea (perhaps not usually expressed so crudely) is very common.... A careful study of the biblical use of the language of conversion, of returning to the Lord, will show that, on the contrary, it is always in the context of concrete decisions at the given historical moment.

10 According to Newbigin, one of the "very practical and indeed painful" questions that conversion brings to a focus is the relation of Christ to our secular history:

> Conversion has always an ethical content; it involves not only joining a new community but also accepting a new pattern of conduct. Conversion implies that the convert accepts this new pattern of conduct as that which is relevant for the doing of God's will and the fulfillment of his reign at this particular juncture of world history. Every conversion is a particular event shaped by the experience of the convert and by the life of the Church as it is at that place and time (Newbigin, *The Finality of Christ*, p. 91).

11 See Richard J. Cassidy, *Jesus, Politics and Society: A Study of Luke's Gospel* (Maryknoll, N.Y.: Orbis Books, 1980). A good summary of relevant material about Jesus's relationship to the authorities can be found in chapter 4, "Jesus and His Political Rulers" (pp. 50–62). Several appendices provide helpful historical background about the Romans and the religious rulers of Jesus's day.

12 See Acts 9:2; 19:9, 23; 22:4, 24:14, 22. While the word *way* is used both literally and figuratively more than one hundred times in the New Testament, its use in an unconditional and absolute sense as a name for the Christian movement is unique to these six passages in Acts. English translations usually capitalize it as "the Way," which becomes a designation for the Christian community and its preaching. G. Ebel, "Way," *NIDNT*, 3:933–947; Wilhelm Michaelis, "Hodos," *TDNT*, 5:42–114.

While the origin of the self-designation has not yet been fully explained, most scholars would agree that the Christian's unique lifestyle contributed to the name. The dictionaries mentioned above emphasize this term as a "designation for Christians and their proclamation of Jesus Christ, which includes the fact that this proclamation also comprises a particular walk or life or way," and refer to "the mode of life which comes to expression in the Christian fellowship."

13 From Aristides, *Apology 15, in The Ante-Nicene Fathers,* ed. Allan Menzies, 5th edition (New York: Charles Scribner's Sons, 1926), 9:263–279.

14 The pagan religions of the day stood in stark contrast to Christian faith both in their separation of belief from behavior and in their refusal to require exclusive loyalty to any one god. As we have seen, biblical conversion demanded a total change of life direction: it forged the vital connection between faith and discipleship, and it called for absolute allegiance to the true and living God. The claim made on the lives of Christian converts was total, unlike the partial and syncretistic observances of the pagan deities. In an atmosphere of lukewarm religious pluralism, the single-minded commitment of Christian conversion "stood out like a sore thumb," says Michael Green. There is a good study of this contrast in the "conversion" chapter (chapter 6) of Green's *Evangelism in the Early Church* (Grand Rapids, Mich.: Eerdmans, 1970): "Helenistic man did not regard ethics as part of religion. . . . This separation of belief from behavior was one of the fundamental differences between the best of pagan

philosophical religion and Christian religion. . . . Conversion, then, in
our sense of an exclusive change of faith, of ethic, of cult was indeed
utterly foreign to the mentality of the Graeco-Roman world"
(p. 146).

  A. D. Nock's classic study, *Conversion* (London: Oxford University
Press, 1933), comes to the same conclusion. Christian conversion
"demanded a new life in a new people" and was a radical change in
behavior as well as belief (p. 7f.). This all-encompassing quality of
conversion is stressed by Nock's contrast between Christian conversion and what he calls general religious "adhesion." Adhesion is the
label-changing kind of conversion that was characteristic of other religious cults.

15 One recent study concludes that the amazing spread of early Chris-
    tianity was due to "a single, over-riding internal factor . . . . the radi-
    cal sense of Christian community—open to all, insistent on absolute
    and exclusive loyalty, and concerned for every aspect of the believer's
    life. From the very beginning, the one distinctive gift of Christianity
    was this sense of community." J. G. Gager, *Kingdom and Community:
    The Social World of Early Christianity* (Englewood Cliffs, N.J.: Prentice
    Hall, 1975), p. 140.

16 Michael Green, *Evangelism in the Early Church*, p. 120.

17 This is quoted from Tertullian's *Apology* 39 in William J. Walsh and
    John P. Langan, "Patristic Social Consciousness: The Church and the
    Poor," *The Faith That Does Justice: Examining the Christian Sources for
    Social Change*, ed. John C. Haughey (New York: Paulist Press, 1977),
    p. 138.

18 Gustavo Gutierrez, *A Theology of Liberation*, trans. and ed. by Sr.
    Caridad Inda and John Eagleson (Maryknoll, N.Y.: Orbis Books,
    1973), p. 205.

CHAPTER 2

 1 Christopher Lasch has written an illuminating and helpful book on
    this topic: *The Culture of Narcissism* (New York: W. W. Norton &
    Co., 1979). He demonstrates in thorough and revealing ways how in-
    fantile narcissism and the American consumer ethic have combined
    to impoverish the quality of our collective life.

 2 Orlando E. Costas, "Conversion as a Complex Experience: A Per-
    sonal Case Study," in John R. W. Stott and Robert Coote, eds.,
    Down to Earth: Studies in Christianity and Culture (Grand Rapids,
    Mich.: Eerdmans, 1980), p. 186.

3 Readers unfamiliar with this biblical insight may want to refer to John H. Yoder's excellent summary in "Christ and the Powers," chapter 8 of his book, *The Politics of Jesus* (Grand Rapids, Mich.: Eerdmans, 1972). The classic beginning work in this area is H. Berkhof's *Christ and the Powers* (Scottdale, Pa.: Herald Press, 1962). William Stringfellow provides good contemporary application about the "traits" and "strategems" of the principalities in chapters 3 and 4 of his book, *An Ethic for Christians and Other Aliens in a Strange Land* (Waco, Tex.: Word, 1973).

4 Baum's essay, "Critical Theology," in chapter 9 of his book, *Religion and Alienation* (New York: Paulist Press, 1975), is one of the best I've seen on the social meaning of sin and salvation. He analyzes in depth the character of collective evil and the social consequences of a gospel message that is purely individual. Both his political and theological insights are on target. This essay is reprinted in an edited version in Walter E. Conn, ed., *Conversion: Perspectives on Personal and Social Transformation* (Staten Island, N.Y.: Alba House, 1978), pp. 281–295.

CHAPTER 3

1 See the November 1978 housing issue of *Sojourners* (7, no. 11) for further information about the speculation that is under way in inner-city neighborhoods. This issue includes "The Housing Crisis Comes Home: Our experiences with displacement of the poor," by Perry Perkins; "The New Refugees: Displacement of the urban poor," by Jim Wallis; "A View from the Industry: How one developer sees the changing city," an interview with Michael Brenneman; and "Signs of Hope in the Cities: Six groups bucking the trend," by Paul Brubaker.

2 See Tom Hanks, "Why People Are Poor: What the Bible Says," *Sojourners* 10, no. 1 (January 1981): 19–22.

3 The profit margins for foreign investments are significantly higher than those for domestic investments by U.S. corporations. When Americans complained about high oil company profits in the 1970s, the companies quickly explained that most of the profits came from overseas. The auto industry crisis of 1980–81 would have come much sooner were it not for billions of dollars in foreign profits. U.S. foreign aid helps to boost these profits by often requiring recipient nations to buy U.S. products. See Richard J. Barnet and Ronald E. Muller, *Global Reach: The Power of Multinational Corporations* (New York: Simon & Schuster, 1974).

4 Quoted in David Kirk, *Quotations from Chairman Jesus* (Springfield, Ill.: Templegate Publishers, 1969), p. 177.

5 The economic analysis presented in this chapter, and many of the statistics cited, come from a variety of sources. The most helpful of these are included in the bibliography for this chapter. A good collection of many of the current statistics and sources can be found in Adam Finnerty's *No More Plastic Jesus: Global Justice and Christian Lifestyle* (Maryknoll, N.Y.: Orbis Books, 1977) and Jack A. Nelson's, *Hunger for Justice: The Politics of Food and Faith* (Maryknoll, N.Y.: Orbis Books, 1980).

6 From John Chrysostom's *On Galatians: Homily* 18, quoted in William J. Walsh and John P. Langan, "Patristic Social Consciousness: The Church and the Poor," in John C. Haughey, ed., *The Faith That Does Justice: Examining the Christian Sources for Social Change* (New York: Paulist Press, 1977), pp. 125–126.

7 Ibid., p. 126.

8 From Cyprian's *The Lapsed*, 11–12, quoted in Walsh and Langan, "Patristic Social Consciousness," p. 123.

9 From John Chrysostom's *On Matthew: Homily* 35,5, quoted in Walsh and Langan, "Patristic Social Consciousness," p. 118.

10 See Donald W. Dayton, *Discovering an Evangelical Heritage* (New York: Harper & Row, 1976). Much of this material appeared in a ten-part series entitled "Recovering a Heritage," published from June/July through May 1975, *Post-American*.

11 David O. Moberg, *The Great Reversal: Evangelism and Social Concern*, Rev. Ed. (New York: J. B. Lippincott Company, 1977, 1972).

12 See John Howard Yoder's "The Implications of the Jubilee" in his *The Politics of Jesus* (Grand Rapids, Mich.: Eerdmans, 1972). This chapter relies heavily on Andre Trocme's *Jesus and the Nonviolent Revolution*, trans. by Michael H. Shank with Marlin E. Miller, (Scottdale, Pa.: Herald Press, 1973). A doctoral dissertation that confirms many of the insights of Yoder and Trocme can be found in R. B. Sloan's *The Favorable Year of the Lord: A Study of Jubilary Theology in the Gospel of Luke* (Austin, Tex.: Schola Press, 1977).

13 Ronald J. Sider, *Rich Christians in an Age of Hunger: A Biblical Study* (New York: Paulist Press, 1977); Bruce C. Birch and Larry L. Rassmussen, *The Predicament of the Prosperous* (Philadelphia: Westminster Press, 1978); Jack A. Nelson, *Hunger for Justice: The Politics of Food and Faith*, (Maryknoll, N.Y.: Orbis Books, 1980); James B. McGinnis,

Bread and Justice: Toward a New International Economic Order (New York: Paulist Press, 1979).

14 One article resulting from our study was "The Bible and the Poor," written by Bob Sabath and published in the February/March 1974 issue of the *Post-American*. A more recent overview of the Old Testament and New Testament teaching about this subject is found in Peter David's two-part article, "God and Mammon," *Sojourners* (February/March, 1978).

15 Quoted in Kirk, *Quotations from Chairman Jesus*, p. 174. Also quoted in Walsh and Langan, "Patristic Social Consciousness," p. 114.

16 From Tertullian's *Apology* 39, quoted in Kirk, *Quotations from Chairman Jesus*, p. 174; and in Walsh and Langan, "Patristic Social Consciousness," p. 138.

17 The *Didache* instructed, "Give to anyone that asks, without looking for any repayment, for it is the Father's pleasure that we should share his gracious bounty with all men." Sharing was to replace possessing as a value in the Christian community. Ambrose, the fourth-century Bishop of Milan, deplored the greed of the rich, defended the rights of the exploited, and regarded the possessions of the wealthy as goods stolen from the poor. "You are not making a gift of your possessions to the poor person," said Ambrose. "You are handing over to him what is his." Basil, a contemporary of Ambrose, said, "The bread in your cupboard *belongs* to the hungry man; the coat hanging unused in your closet *belongs* to the man who needs it; the shoes rotting in your closet *belong* to the man who has no shoes; the money which you put in the bank *belongs* to the poor. You do wrong to everyone you could help but fail to help."

John Chrysostom spoke against legalized and institutionalized injustice: "The rich are in the possession of the goods of the poor, even if they have acquired them honestly or inherited them legally." The Constantinople church father said that, if the rich do not share, then they are "a species of bandit." Chrysostom's attack on the rich is based on the patristic principle of equality. "Do not say, 'I am using what belongs to me.' You are using what belongs to others," said Chrysostom. "All the wealth of the world belongs to you and to others in common, as the sun, air, earth, and all the rest."

For readers who want to pursue further the teachings of the early church fathers, I recommend two sources. The first is an essay, "Patristic Social Consciousness—The Church and the Poor," by William

J. Walsh, S.J., and John P. Langan, S.J., published in *The Faith That Does Justice: Examining the Christian Sources for Social Change*, edited by John G. Haughey (New York: Paulist Press, 1977). The second good source is a humorously titled book by David Kirk, *Quotations from Chairman Jesus* (Springfield, Ill.: Templegate Publishers, 1969); see the chapter called "Fathers of the Revolution." All the statements quoted here can be found in these two sources.

18 Bread for the World, "World Hunger and Poverty," www .bread.org/ hungerbasics/international.html

19 From John Chrysostom's *On Matthew: Homily* 48.8, quoted in Walsh and Langan, "Patristic Social Consciousness," p. 130.

20 Donald W. Dayton, *Discovering an Evangelical Heritage* (New York: Harper & Row, 1976), p. 15. See chapter 2, "Reform in the Life & Thought of Evangelist Charles G. Finney," pp. 15–24.

21 Ibid., p. 18f. All of the Finney quotes in these two paragraphs are also from this chapter.

CHAPTER 4

1 Richard B. Hays, "A Season for Repentance: An Open Letter to United Methodists," *The Christian Century*, August 24, 2004.

2 Dietrich Bonhoeffer, *The Cost of Discipleship*, Rev. Ed. (New York: Macmillan Publishing Co., 1974, 1937 first published), p. 99. See the chapter entitled "Discipleship and the Cross."

3 Glen Stassen, "*Just Peacemaking: Ten Practices for Abolishing War*," Summary of the theory for the panel on Just Peacemaking Theory at the annual meeting of the American Academy of Religion, Orlando, Florida, November 1998, www.fullerseminary.net/sot/faculty/stassen /cp_content/homepage/homepage.htm

CHAPTER 5

1 A helpful discussion of community and its relationship to conversion, social change, and the overall plan of God for reconciliation can be found in *The Spirit and the World* (New York: Hawthorne Books, 1975) by James W. Jones, professor of religion at Rutgers College.

2 See Jean Vanier, *Community and Growth: Our Pilgrimage Together* (New York: Paulist Press, 1979), p. 5. "A community is only a community when the majority of its members is making the transition from 'the community for myself' to 'myself for the community.'"

CHAPTER 6

1 Karl Barth, "The Holy Spirit and the Upbuilding of the Christian Community," *Church Dogmatics*, IV/2 (Section 67): 638f. Some of the ideas in this and the next several paragraphs are adapted from a previous article by Jim Wallis and Robert Sabath, "The Spirit in the Church," *Post-American* 4, no. 2 (February 1975): 3–5.

2 See John Howard Yoder, "Binding and Loosing," *Concern No. 14: A Pamphlet Series for Questions of Christian Renewal* (February 1967). Currently out of print, and back issues unavailable.

3 Oscar Cullmann, *Christ and Time: The Primitive Christian Conception of Time and History*, trans. Floyd V. Filson (Philadelphia: Westminster Press, 1950), p. 228.

4 See Wallis and Sabath, "The Spirit in the Church," and Yoder, "Binding and Loosing," both cited above.

5 For an early history of the development of Sojourners community, see "Crucible of Community: A dialogue on the shaping of Sojourners," *Sojourners* 6, no. 1 (January 1977): 14–21.